OLD YEASTS - NEW QUESTIONS

Edited by **Cândida Lucas** and **Célia Pais**

Old Yeasts - New Questions
http://dx.doi.org/10.5772/66616
Edited by Cândida Lucas and Célia Pais

Contributors

Marcos Antonio De Oliveira, Carlos Breyer, Leonardo Schultz, Carlos Tairum Jr, Melina Dos Santos, Anderson Da Cunha, Karen Romanello, Z. Petek Cakar, Burcu Turanli-Yildiz, Burcu Hacisalihoglu, Enas Sayed, Mohammad Abdelkareem, Dorota Kregiel, Antonio Peña, Martha Calahorra, Norma Silvia Sánchez, Esperanza Del Pilar Infante Luna, Ileana Farcasanu, Lavinia Ruta, María Esperanza Cerdán, Ángel Vizoso-Vázquez, Aida Barreiro Alonso, Agustín Rico Diaz, Mónica Lamas-Maceiras, Esther Rodríguez-Belmonte, Manuel Becerra-Fernández, María-Isabel González-Siso

Published by InTech

Janeza Trdine 9, 51000 Rijeka, Croatia

Publishing Process Manager Romina Rovan
Technical Editor SPi Global
Cover InTech Design team

First published December, 2017
Printed in Croatia
Legal deposit, Croatia: National and University Library in Zagreb

Additional hard copies can be obtained from orders@intechopen.com

Old Yeasts - New Questions, Edited by Cândida Lucas and Célia Pais
p. cm.
Print ISBN 978-953-51-3677-4
Online ISBN 978-953-51-3678-1

For EU product safety concerns:
IN TECH d.o.o., Prolaz Marije Krucifikse Kozulić 3, 51000 Rijeka, Croatia,
info@intechopen.com or visit our website at intechopen.com.

Meet the editors

Cândida Lucas is a Portuguese, born at Porto in 1958. She has a PhD degree in Biology/Microbiology from the New University of Lisbon/Gulbenkian Institute of Science (1988), following a degree in Biology (University of Lisbon [1982]). She is a full professor of the Biology Department from the University of Minho, Portugal, since 2015. Then, since 1990, she contributed to create postgraduation courses and programmes on Molecular Biology/Genetics. She directed the Centre of Molecular and Environmental Biology Research (CBMA) (2006–2013) and is presently a codirector of the Institute of Science and Innovation for Bio-Sustainability (IB-S) from the University of Minho (http://ib-s.uminho.pt/). Her scientific interests covered diverse aspects of yeast biology, from plasma membrane transporters and osmotolerance to the extracellular matrix and differentiation (h-index 18).

Célia Pais has a PhD degree in Biology/Microbiology from the University of Minho, Portugal (1990), and a postgraduate degree in Plant Pathology from the Imperial College, London, UK, following a degree in Biology at the University of Coimbra, Portugal (1978). She is an associate professor with habilitation at the Biology Department, University of Minho, where she was the vice-president of the School of Sciences (2002–2006) and the head of the Biology Department (2012–2014). She participated in the creation of the PhD Programme in Environmental and Molecular Biology that she coordinated from 2008 to 2012. She develops her scientific activity at the Centre of Molecular and Environmental Biology Research (CBMA), focusing on the study of yeast diversity and the development of new molecular tools for diagnosis and genotyping.

Contents

Preface

Yeast-based biotechnology traditionally regards the empirical production of fermented drinks and leavened bread. Although these processes are known to man for millennia, they keep posing challenges and fuelling considerable amount of research. Still, feasible and economically viable applications of yeast-driven technology are presently very diverse, critically contributing to the economical welfare of many differently developed countries and representing at world scale a business volume worth billions.

Yeast biotechnology follows essentially two main strategies. One uses the organism as a cell factory for the efficient production of fine chemicals and molecules to apply on pharma or research or bulk chemicals as in the case of biofuels. Another strategy uses the organism as bio-tool applied *in loco* to execute specific tasks. Examples can be food preservation implicating on shelf-life span, the concentrative removal of metals from industrial or mining residues or bioremediation of environmental disasters like oil spillage in the sea.

Yeasts are a very large group of microorganisms, harbouring a large number of species, and are genetically very diverse. Like bacteria, they are present in all types of ecosystems, including human microbiome. Still, derived from the traditional applications abovementioned, most of the yeast biotechnologies of today are still based on the utilization of *Saccharomyces cerevisiae* strains, either wild-type isolates from specific environmental niches or engineered, be it genetically, metabolically or evolutionarily. The possibilities associated with yeast biodiversity have therefore been largely disregarded.

The big disproportion between the numbers of applications that use *S. cerevisiae* in relation to other species has other reasons besides tradition. Successful biotechnology depends on detailed knowledge, not only of the process but also mainly of the organism that governs it. *S. cerevisiae* is by far the most well-studied yeast species and one of the best well-known organisms, maybe just overcome by *Escherichia coli*. It is easier and cheaper to research comparing to higher eukaryote models and does not pose manipulation ethical problems or health concerns. Importantly, yeasts are recognized as excellent models of cell and molecular biology for higher eukaryotes, including humans. Not only do yeasts contribute with key discoveries to understand cellular processes, but they also contribute with crucial knowledge about complex diseases, some of which can only be studied in yeast, like mitochondrial metabolic diseases or neurodegenerative disorders like Batten or Huntington diseases.

Yeast diversity of species and strains constitutes per se a huge potential for biotechnology. It though requires considerable reliability, which is only obtained with more research and knowledge. Many species have been poorly addressed by research and are consequently less applied, most of them not at all. The general designation of the non-*Saccharomyces* yeasts as *non-conventional* well translates this disproportion. Two exceptions can be pinpointed. *Schizosaccharomyces pombe*, the fission yeast, has been largely used as a model to understand cell cycle control. *Candida* species have been largely used to address host defences against pathogens and yeast-promoted tissue invasion and infection. This last case has gained particular importance in view of the increasing severity and prevalence of human yeast infections.

In spite of the huge potential that yeasts present for biotechnology and health, research with yeasts has presently entered a stalling period, yeasts being delegated to non-mainstream scientific interests difficult to fund. Of course, much has been achieved in the last 50 years, but yeasts, including *S. cerevisiae*, still present many challenges and still can surprise by unveiling unprecedented scientific knowledge, opening new strands for application and innovation. Moreover, it is not only the organism itself that appears to have become *old fashion*. Research in the last 20 years has focused essentially on molecular aspects. This was fuelled by the growing amenability of yeast genetic manipulation and availability of molecular tools. This enabled the building up of a robust pile of knowledge on signalling, transcription and post-translational modification and regulation, trafficking, chromosomal structure and behaviour and so on and so forth. But the actual function of the yeast cell transcends the mechanistic vision of intertwining enzymatic production lines that we can manipulate at our will to yield interesting metabolites for our profit. There are knurls, kinks and nuts. Critically, new and old unanswered questions cloud a full-picture understanding of yeast biology and preclude biotechnology application success. It is necessary to not lose the ability to assess biochemical and biophysical aspects in order to really understand function.

This book provides some insights into aspects of yeast science and yeast-based biotechnology less frequently addressed in the literature but nonetheless decisive to improve knowledge and, accordingly, boost up yeast-based innovation. These generally regard how the yeast cells dispose of ions, implicating in how they control the electric plasma membrane potential (Chapter 1), in how they act as biosorbents of heavy metal ions without suffering from their toxicity (Chapter 2) and their ability to transfer electrons as a whole cell (Chapter 3). All of these have clear technological implications in the use of yeasts in bioremediation, biomining or the development of yeast-based fuel cells. All these processes are addressed as expected in *S. cerevisiae*, in which metabolic engineering for the specific production of certain dine and bulk chemicals is reviewed (Chapter 4). But also the biotechnological potential of several so-called non-conventional yeast species is briefly evaluated (Chapter 5), for the creation of yeast-based microbial fuel cells (Chapter 3) as well as for fermentative processes, either alone or in mixed fermentations (Chapter 5). Moreover, the mechanisms and molecules involved in cellular relief from hydroperoxides, which participate in tissue injury and in the onset and progression of degenerative diseases in humans (Chapter 6). These mechanisms are crucial to understand ageing, as are the mechanisms involved in DNA protection (Chapter 7), needless to stress how much these studies are crucial for pharmacological and clinical development. Finally, the last but an important chapter (Chapter 8) presents a whole new view on how yeasts can also colonize plants, living as endophytic microorganisms. This could parallel the commensality of yeasts in the human tissues. Yet, too few information is still available to know whether in plants, like in mammals, yeasts can change drastically their biology, shifting commensalism into parasitism and developing an infection. This is a quite promising line of work, bearing in mind the growing worldwide agriculture problem posed by plagues that either resist chemical treatment or cannot be managed that way because of health-risk concerns. New and revolutionary strategies are needed to control yeast and fungi pathogens in plants as in animals, which can only be achieved through continuous research effort.

Cândida Lucas
Full Professor
Institute of Science and Innovation for Bio-Sustainability (IB-S)
Centre of Molecular and Environmental Biology Research (CBMA)
University of Minho, Portugal

Célia Pais
Associated Professor
Centre of Molecular and Environmental Biology Research (CBMA)
Biology Department, University of Minho, Portugal

Section 1

Yeasts - Indispensible Biotech Playwers

Chapter 1

The Plasma Membrane Electric Potential in Yeast: Probes, Results, Problems, and Solutions: A New Application of an Old Dye?

Antonio Peña, Norma Silvia Sánchez and
Martha Calahorra

Additional information is available at the end of the chapter

http://dx.doi.org/10.5772/intechopen.70403

Abstract

For a long time, estimations and actual measurements of the electric plasma membrane potential (PMP) in whole yeast cells have been the subject of studies by several groups without reliable results. The conditions in the measurements, as well as precautions required to perform them, are described here. Essentially, two approaches using different dyes are reviewed: (a) qualitative estimations by following fluorescence changes under different energization conditions and (b) measurements of the PMP by the accumulation of dyes. An analysis is presented regarding the conditions recommended to obtain more consistent results when following the fluorescence changes. Also, measurements of accumulation of different dyes, and the necessary conditions to perform them, are analyzed. In particular, using acridine yellow appears to be a trustworthy method, with few reserves, both to follow in real time the qualitative changes of the PMP by fluorescence changes and to assess actual PMP values by measuring the accumulation of the dye.

Keywords: plasma membrane potential, yeast, acridine yellow, fluorescent monitors

1. Introduction

1.1. Qualitative estimation of the electric plasma membrane potential (PMP) in yeast

After the original proposal for the mechanism of K^+ transport in yeast [1], it was shown [2, 3] that this ion is transported because a H^+-ATPase exists in the plasma membrane, pumping protons outside, therein generating an electric membrane potential difference (PMP), negative inside. This is then responsible mainly for the uptake of cations and other molecules required by the cells through different active uptake systems of which the most prominent is Trk1p [4]. This mechanism was also described for *Neurospora crassa* [5] and many other fungi and also for plants. These findings gave rise to a series of attempts to follow changes in the plasma membrane potential (PMP) in whole yeast cells, initially by measuring the accumulation of cationic compounds [6, 7]. Results were though far from satisfactory, mainly because of the slow rate at which these molecules entered the cells.

The use of fluorescent indicators originally to estimate the membrane potential in mitochondria started a long time ago [8, 9] by observing ethidium bromide fluorescence quenching under different energy states. Also, measurements of the PMP in animal cells were performed with reasonable results [10–12]. Following these studies and using whole yeast cells, fluorescence

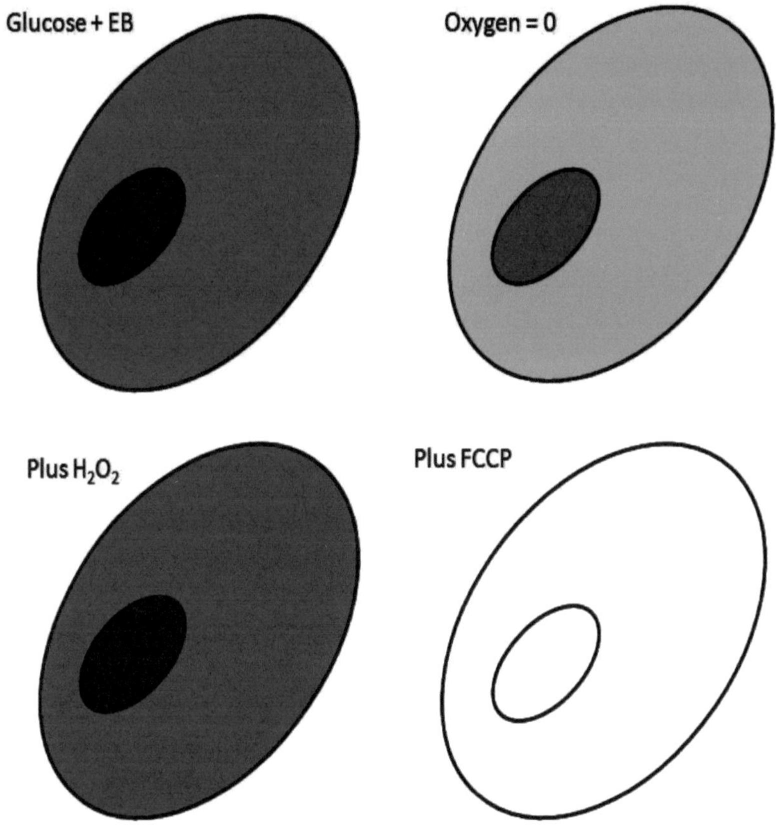

Figure 1. Graphic interpretation of the fluorescence changes of ethidium bromide in yeast under different conditions [13]. (A) When glucose alone is added, fluorescence of the dye is moderately increased, but a large part of it is accumulated by the mitochondria, where its fluorescence is quenched. (B) When oxygen is exhausted, mitochondria are partially de-energized (they still can maintain most of its membrane potential by using ATP generated in glycolysis). (C) Upon the addition of H_2O_2, mitochondrial membrane potential is fully recovered, resulting in fluorescence quenching. (D) The addition of an uncoupler (FCCP) produces the full collapse of the mitochondrial membrane potential, the uniform distribution of the dye in the cytoplasm, and a large increase of fluorescence. The small oval shape represents a mitochondrion.

changes of ethidium bromide under different energization conditions were studied [13]. It was found that in order to observe the fluorescence increase derived from the accumulation of the dye, it was more convenient to use starved yeast cells, which required the addition of glucose as a substrate. **Figure 1** shows a schematic representation of the results observed and their interpretation. It was found that in the presence of glucose, fluorescence showed a rather small increase, and after a few seconds was followed by another one, coincident with the oxygen exhaustion in the medium that as expected could be reversed by the addition of H_2O_2. This led us to suspect that the initial fluorescence increase was actually composed of not only an increase due to the accumulation of the dye in the cytoplasm but also a decrease due to its accumulation in mitochondria, according to what had been described before to what was happening in the isolated mitochondria [9]. Moreover, it was found that, also as expected, the addition of an uncoupler resulted in a much larger increase of fluorescence, consistent with the idea that collapsing the membrane potential of the mitochondria resulted in an efflux of the dye and its uniform distribution in the cytoplasm, where the dye was highly fluorescent [14]. The actual fluorescence changes observed in the spectrofluorometer were similar to those shown in **Figure 2** with acridine yellow.

Several years later, a new procedure, under the same principles was proposed by following the fluorescence changes of the cyanine $DiSC_3(3)$ [14]. Since then, perhaps the most used monitor to follow changes of the PMP in yeast is $DiSC_3(3)$, a high affinity probe that in the presence of a substrate (glucose) is concentrated in the cytoplasm, and similarly to ethidium bromide, also in the mitochondria [16]. Although the latter can be avoided by using low concentrations

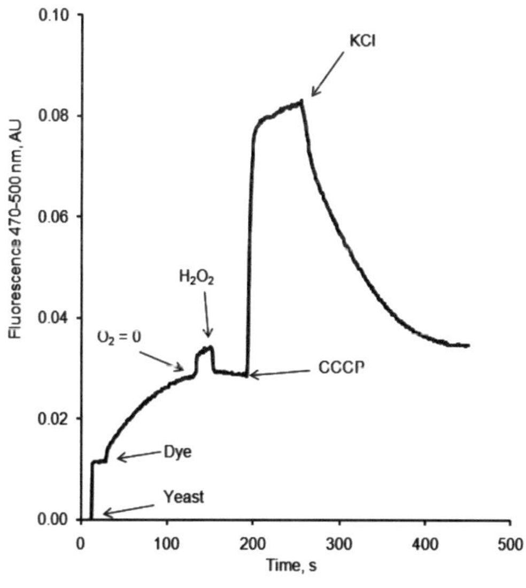

Figure 2. Fluorescence changes of 25 μM acridine yellow at 470–500 nm when added to *S. cerevisiae* cells. Incubation was carried out in 10 mM MES-TEA buffer, pH 6.0; 10 μM $BaCl_2$; 20 mM glucose in a final volume of 2.0 mL. Where indicated, 8 μL of 3% H_2O_2, 10 μM CCCP or 5 mM KCl were added. Representative figure of one experiment similar to that already reported [15]. Fluorescence was followed in an SLM Aminco spectrofluorometer with continuous stirring and at a temperature of 30°C. Excitation and emission fluorescence were 470 and 500 nm, respectively. Phototube voltage was 600 V, and slits were set at 8 nm. AU, arbitrary units.

of an uncoupler, still the high affinity and binding of the dyes to the cell components turns it difficult to calculate their free concentration inside, in order to accurately measure the PMP through its accumulation. Other authors [17] also used $DiOC_6(3)$ with this same purpose with rather uncertain values. Thus, results using cationic molecules, mostly dyes, have been under discussion for many years without arriving to a general consensus about the methods, and even less regarding the results obtained, especially when it comes to the real values of the PMP in yeast.

2. The quantitative measurement of the PMP

$DiSC_3(3)$ has been one of the most used monitors to estimate changes of the PMP in yeast [14, 16, 18–22] by following its fluorescence changes under different energy conditions of yeast cells. Attempts have also been made to assess the actual PMP values from the internal and external concentration ratios between the cells and the media. This implies (a) measuring the concentration and the amount of the dye or any cationic agent remaining after incubation with the cells; (b) knowing the amount originally added that taken up by the cells can be obtained; (c) the internal concentration of the dye can be obtained by previously measuring the internal water volume of the cells to acquire the internal concentration. Finally, by using the Nernst equation, the value of the PMP, in millivolts can be obtained from:

$$E = -(RT/ZF)\ln C_{in}/C_{out} \tag{1}$$

where R is the gas constant, Z the charge of the ion, F is the Faraday constant, C_{in} and C_{out} the internal and external dye concentrations. Then, PMP is approximately equal to:

$$PMP = -60\left(\log\,[C_{in}]/[C_{out}]\right). \tag{2}$$

In this way, approximate values of the PMP have been obtained [14, 15, 17]. However, those values are subject to many errors and uncertainties that will be discussed below.

Studies with *Saccharomyces cerevisiae* [18] and *Rhodotorula glutinis* [19] proposed another approach to measure the actual PMP by following the changes of the λ_{max} of the fluorescence spectrum of this monitor as an indicator, which has been also used by other authors [20, 23]. Within the claims to actually measure the plasma membrane potential in yeast, some values reported are too low to explain, among others, the large accumulation of K^+, which can reach internal concentrations of around 300 mM against micromolar external concentrations of the cation. Results are hard to rationalize, even considering that one part of the cation redistributes into the vacuole [24] and another is neutralized by the accumulation of bicarbonate when glucose is the substrate [25]. The use of fluorescent probes not only for the estimation of plasma membrane potential but also for many other purposes was partially reviewed by Slavik [26], and more recently, for yeast cells [16, 19]. However, this topic continues to be unsolved, and it is our belief that several aspects should be considered. The problems to estimate and measure the PMP in yeast imply a large series of factors that may affect results, such as the following: (a) the influence of the binding of the cationic monitor to the surface of the cell; (b) the use of an adequate buffer, avoiding organic molecules and other cations that may be taken up by the cells; (c) the concentration of the dye, critical and different for each one and probably for different yeast strains or species to observe the fluorescence changes and accumulation; (d) the accumulation of the dye by the mitochondria; (e) the binding of the dye inside the cells, and (f) the use of starved cells that allows observing changes due to energization of the cells by adding a substrate.

3. Interaction, uptake, distribution, and binding inside the cells

3.1. Binding to the surface

The first interaction of the dyes is, of course, with the negatively charged cell surface. Since the first studies performed [14], an immediate increase of fluorescence of $DiSC_3(3)$ was observed upon its interaction with the cells that could be diminished by the addition of low concentrations of a divalent cation eventually recommending the use of $BaCl_2$ to avoid binding [16].

3.2. Uptake

Using starved cells, the addition of a substrate, usually glucose, is required in order to generate the PMP [14]. Most probes appear to enter the cells by free diffusion. However, ethidium bromide, at least at certain concentrations, seems to be transported into the cell through the K^+ transport system [27]. Monitors are generally cationic, with a delocalized electron structure that nonetheless does not eliminate their positive charge. In general, anionic molecules do not seem to enter the cells [28]. The main relevant characteristic to the topic of this review is that cationic molecules employed seem to be driven inside by the plasma membrane electric potential difference, and because of this, they can be used to follow changes of the membrane potential under varied conditions [16, 18, 19, 21, 22].

3.3. The internal distribution

Once the dyes enter the cells, they are not uniformly distributed inside due to the negative inside membrane electric potential difference of the mitochondria. Experiments show [16] that in fact the changes observed in the dyes' fluorescence within yeast cells mainly when using ethidium bromide, $DiSC_3(3)$, and acridine yellow [15] depend on the addition of a substrate, generally glucose. When this is added (a) a slow increase of fluorescence is seen, then after a few seconds that is followed by another small increase, coincident with the exhaustion of oxygen that can be reversed by the addition of a small concentration of H_2O_2, (b) if then a low concentration of an uncoupler around 5–15 μM is added, such as CCCP or FCCP, a much larger increase of fluorescence is observed. Finally, (c) when a concentration of KCl enough to be transported inside is added, a decrease of the fluorescence is seen. One additional observation with at least two dyes, $DiSC_3(3)$ [15, 16] and acridine yellow is that they under none of these conditions enter the vacuole. **Figure 2** shows the results of one experiment in which these fluorescence changes were observed with acridine yellow but are similar to those observed with $DISC_3(3)$ or ethidium bromide.

3.4. Interpretation of results

Results shown in **Figure 2** are interpreted as follows: in the presence of a substrate necessary to energize the cells, the dye is transported inside, driven by the PMP. Once inside, it is also taken by the mitochondria, where a large accumulation takes place, resulting in quenching of most of the fluorescence. This observation is supported by the small fluorescence increase when oxygen in the medium is exhausted by respiration that can be reversed by the addition of H_2O_2. When oxygen is exhausted, a partial deenergization of mitochondria occurs. Both these changes are small because even in the absence of oxygen, mitochondria are energized by the ATP produced in glycolysis. Adding then a low concentration (5–15 μM) of CCCP or FCCP depolarizes mitochondria, producing the efflux of the dye from the organelles and its uniform distribution in the cytoplasm, resulting in a large fluorescence increase (dequenching). After this, K^+ addition, since the ion must be taken up through a PMP-driven transport partially neutralizing it, produces a large decrease of the fluorescence. Another important result in this respect is that the addition of a similar concentration of NaCl does not produce the fluorescence decrease resulting from the addition of KCl because Na^+ affinity for the transporter is much lower than that of K^+(not shown).

3.5. Binding to the internal cell components

One important characteristic of the monitors employed is the low fluorescence they show only in the buffered medium, which is largely increased by their interaction with the internal components of the cell [14, 15]. This characteristic is actually the basis for the fluorescence-based studies of the PMP and its changes under the different conditions. However, knowing the amount bound to the internal components of the cells becomes one of the main problems when trying to measure the actual values of the PMP by their accumulation.

4. Evidence from microscope images

Additional evidence is provided by looking at the cells under the microscope. **Figure 3** shows the images obtained with acridine yellow in a similar experiment to one already reported [15]. It can be clearly seen that in the presence of glucose alone the dye is concentrated in the mitochondria, but fluorescence is rather faint, which confirms that in fact, most of the dye is accumulated by these organelles; its fluorescence is quenched because of its large accumulation (**Figure 3A**). Then, particularly with acridine yellow, upon the addition of a low concentration of an uncoupler (10 μM CCCP), fluorescence of the dye increases remarkably and gets uniformly distributed in the cytoplasm (**Figure 3B**). The subsequent addition of KCl results in a general decrease of the fluorescence (**Figure 3C**). It is important to point out that the dye does not enter the vacuole.

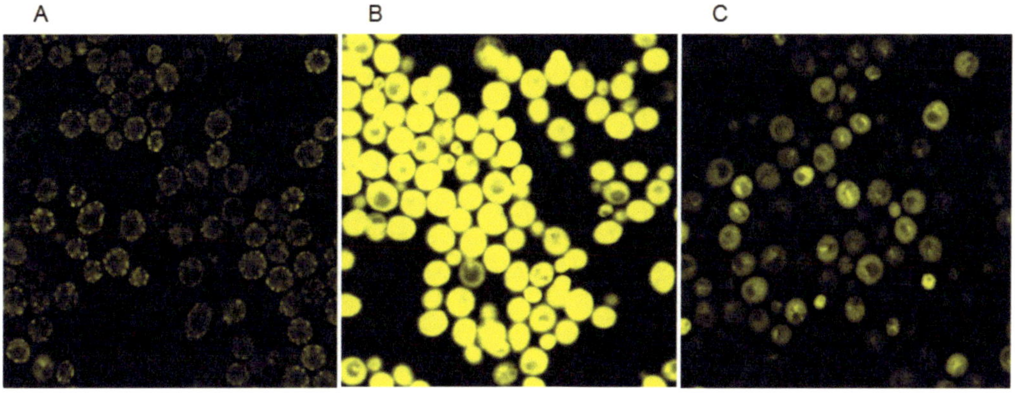

Figure 3. Microscopic images following fluorescence changes of acridine yellow in yeast as a result of differences in PMP. In (A), the cells were incubated in 10 mM MES-TEA, pH 6.0, 20 mM glucose and 50 μM acridine yellow and observed 5 min later. Then, 10 μM CCCP was added, and 5 min later, image B was obtained. Image C was obtained 5 min after adding 10 mM KCl to the same preparation of image B. Similar figure to that of [15].

5. The problems and solutions tested

The general procedure to follow the fluorescence changes is as follows. The cells are added to an adequate buffer, and as a substrate, 20 mM glucose is used. A 2.0 mL final volume of the medium is important to ensure the effective continuous mixing of the incubation mixture. Adequate settings of the instrument regarding high voltage applied to the photomultiplier and the slit width should be chosen to get an adequate level of the fluorescence signal and its changes. We use an SLM Aminco spectrofluorometer with a cell holder provided with continuous magnetic stirrer and a constant temperature system. Different factors affecting the values obtained are the following:

Buffer. Salts that may be taken up by the yeast cells such as phosphate, sodium, or potassium salts should not be used to avoid their interference with the functioning of the cells. We prefer 10 mM MES (morpholinoethanesulfonic acid) taken to pH 6.0 with triehanolamine.

Amounts of cells and dye. With different monitors, different amounts of cells and dye concentrations should be assayed to find those in which the best tracings are obtained.

External binding. When fluorescence is followed incubating the cells in the presence of an adequate buffer, even in the absence of a substrate, an increase of fluorescence is observed, which appears to be due to the simple binding of the dyes to the negative external charges of the membrane. To avoid this, we started using a low concentration of $CaCl_2$ [14], but later realized that it was better to use a low concentration of 10–20 μM $BaCl_2$ [16].

Accumulation in the mitochondria. Uncouplers in yeast, at least *S. cerevisiae*, show an interesting behavior. They produce a very clear stimulation of respiration at concentrations around 2–5 μM, while to inhibit K^+ transport, a process dependent on PMP, several times higher concentrations are required [16, 27]. This indicates that low concentrations can uncouple the mitochondria, without affecting the PMP. The solution then is to use low concentrations (5–15 μM) of either CCCP or FCCP to eliminate the accumulation of the dyes by the mitochondria.

Binding to the internal components. This factor is impossible to eliminate, and it is the actual basis of the analysis of the PMP by fluorescence, since as already mentioned, the monitors used in these studies show very low fluorescence values in water and require their interaction with the internal components of the cells. However, as it will be seen ahead, binding values can be measured.

6. Conclusions from the fluorescence changes

Changes of fluorescence of different monitors are an excellent way to follow not only the qualitative variations of the PMP in yeast, depending on the dye and conditions used. This has been shown with ethidium bromide [13, 27]. Regarding the claims that follow the displacement of the maximal fluorescence, peaks can provide the way to determine the actual value of the PMP and should be taken with caution [19]. Two reasons allow to affirm this (a) as mentioned before, the values obtained are too low to explain the large accumulation of K^+ of around 300 mM inside the cells in the presence of micromolar concentrations outside, even considering that part of it is within the vacuole [24], or partially neutralized by the simultaneous accumulation of bicarbonate [16], and (b) although the authors state that the PMP can be fully collapsed by the addition of 10–20 μM of the usual uncouplers, CCCP or FCCP, results show that the concentrations required to stimulate respiration are 5–10 μM and those to inhibit K^+ uptake are more than 20 μM [16]. Moreover, this anomalous incapacity of uncouplers to freely transport H^+ through the yeast plasma membrane has been used to measure the internal pH of yeast cells by following the distribution of 2,4-dinitrophenol [2, 29]. When using acridine yellow [15], the fluorescence changes observed when deenergizing the mitochondria are particularly large. This means that this dye may be particularly useful to monitor the changes of the electric potential difference of mitochondria.

7. Actual measurement of the PMP by the probe accumulation

Initial attempts to measure the PMP based on the accumulation of cationic agents [6, 7] were unreliable because of the slow entrance of the cationic agents used and the apparently incomplete equilibrium reached between the inside and outside of the cells. Previous calculations with the simple accumulation of different dyes, either ethidium bromide [13], or $DiSC_3(3)$ [14, 16], gave results that appeared too high, and the variations under the different conditions too small. As already pointed out, more recently [15], acridine yellow was found to require higher concentrations than $DiSC_3(3)$ to observe the fluorescence changes usually described (**Figure 2**). Its behavior was also similar to that reported for other indicators when observed under the microscope (**Figure 3**). In summary, its activity was typical of that observed with other dyes, such as ethidium bromide or $DiSC_3(3)$. One difference is that while $DiSC_3(3)$ had to be used at nM concentrations, this dye required a concentration of 50 or even 150 μM to clearly observe changes in its accumulation.

7.1. The PMP measured by the accumulation of acridine yellow

Following a similar procedure to that used with other dyes, its accumulation was measured to calculate the PMP under different conditions. In these experiments, from the amount of dye remaining in the supernatant after centrifuging the cells, we could calculate its internal amount, and from the value of the internal water content, its internal concentration, which allowed to calculate the apparent noncorrected Δψ values that are shown in **Table 1**.

These results were already encouraging and different to those obtained before with $DiSC_3(3)$ [16], but we still had to consider that at least part of the dye was not free, but bound to the internal components of the cells. Values were also interesting regarding their magnitude and reproducibility. With glucose alone, the highest accumulation was observed present in both the cytoplasm and the mitochondria. CCCP addition produced a decrease because the dye was no longer concentrated in the mitochondria, reaching a new equilibrium with the external concentration. As expected, adding 5 mM KCl, which should at least partially collapse the PMP, produced another large decrease. It is also important to emphasize that the addition of NaCl produced only a small change.

7.2. Binding of the dye inside the cells, a possible solution

Considering that we had no way to produce the efflux of the dye bound to the internal components of the cell, we decided to use another approach, trying to measure it. Chitosan, a cationic polymer mostly composed of glucosamine is very effective to permeabilize the plasma membrane of *Candida albicans* [30], but actually this effect was found before on *S. cerevisiae* (unpublished). We therefore used the cells incubated with glucose plus CCCP and added to

	Δψ, mV ± std. dev
Glucose	−168.4 ± 7.3
CCCP	−105.2 ± 1.8
CCCP + KCl	−46.0 ± 4.5
CCCP + NaCl	−95.7

Experiments were conducted using the typical incubation mixture with 10 mM MES-TEA buffer, pH 6.0, 10 μM $BaCl_2$, and 20 mM glucose, with 50 μM acridine yellow and 250 mg of cells (wet weight) in a final volume of 10.0 mL. Absorbance spectra of the samples, and the mean readings between 409 and 411 nm were obtained. From these readings and the linear part of a standard curve, the external concentrations were calculated. Accumulated dye was measured 5 min after the indicated successive additions by determining the concentration, and from it the amount of dye in the supernatant after centrifuging the cells. The internal amount of dye was obtained by subtracting that amount from that originally added. Its concentration was calculated considering the value of internal water of yeast, which has been measured and equivalent to 0.47 mL g^{-1}, wet weight (33). The values were obtained in each case 5 min after the successive addition of (a) cells; (b) CCCP (10 μM), and (c) either KCl or NaCl (5 mM). Δψ was calculated from the Nernst equation, considering the log of the quotient of the internal/external concentrations.

Table 1. Raw calculations of Δψ from the measured accumulation of acridine yellow in yeast.

them 100 μg of chitosan. After centrifuging the cells, it was possible to calculate the amount of the dye leaking out, and of course, that remaining inside bound to the internal components. In this way the amount of dye that remained in the cells after the addition of chitosan could be subtracted from that remaining under the different conditions. Results of one typical of those experiments are the following:

Chitosan produces the permeabilization of the plasma membrane of the cells, producing the efflux of the dye, but still part of it remains inside, bound by its cationic nature. This is an interesting approach to subtract the contribution of the internal binding of the dye in the calculations of $\Delta\psi$, at least in this yeast. However, still one problem exists: the values of the dye remaining inside the cells when KCl was added were found to be lower than those obtained with chitosan. This may be because K^+ may produce an additional displacement of the also cationic dye from its binding sites.

7.3. The exclusion of the dye from the vacuole

Results from the accumulation of the dye considered a uniform distribution inside the cells, whose total water content has been measured and estimated at 47% of water per g of wet weight [31]. Microscope images show that after the uncoupler, the dye is no longer in the mitochondria and distributes in the cytoplasm, but it is absent from the vacuole [15]. This implies that the distribution volume of the dye is smaller than that in the total cell water. Considering this, the internal water in which the dye distributes is not the total value of 0.47 mL g^{-1} [31], but 0.355 mL g^{-1} of cytoplasmic water, excluding the vacuole. Using this new volume, the values shown in **Table 2** were obtained, indicating a still higher value of the PMP, shown in parenthesis in **Table 2**.

	[Int] nmoles/mL	[Ext] nmoles/mL	Ratio	Log	mV
Control	1953	4.1	476	2.67	−161
CCCP	1244	20.8	60	1.77	−107
KCl	284	64.4	4.4	0.64	−39
NaCl	1023	25.9	39	1.60	−96
Chitosan	391	40.4			
Correction for internal binding					
Control	1562	4.1	380	2.6	−155 (211)
CCCP	853	20.8	41	1.6	−97 (196)
KCl	0.7	43.3	?	?	?

The experiment was performed as described in **Table 1**, but where indicated, 100 μg of chitosan was included, and the last sample was taken 5 min later. The amount of the dye that had not entered the cells, subtracted from the total added, allowed to get the amount that entered the cells. To obtain the internal concentration of the dye, this latter amount was divided by the internal water content of the cells, which was estimated before for this yeast strain of 0.47 mL g^{-1} of cells (33). From the internal and external concentrations, the quotient and its log were obtained, and using the Nernst equation, $\Delta\psi$ was calculated. Values in parenthesis were obtained considering that if the dye does not enter the vacuole, the total water in which the dye was distributed was only 0.355 mL g^{-1}. Question marks indicate that the values were not calculated because the efflux of the dye with KCl was larger than that with chitosan.

Table 2. Values of the accumulation of acridine yellow obtained in a representative experiment under different conditions.

8. Concluding remarks

Our work is a constant attempt to clarify a long time controversy, first about the estimation of the fluorescence changes of different monitors with different methods. We could add to the list acridine yellow, long known, but also quite inexpensive, which can be used to that purpose. With this dye larger concentrations are required, which may be probably due to the fact that it has a lower binding affinity to the internal components of the cells. This can be inferred already by comparing the results of its uncorrected uptake values as shown in **Table 3**.

With these corrections, larger differences under the conditions tested could be obtained with acridine yellow, as compared to those obtained with other agents, but much higher than those reported by other authors [19]. The simple accumulation in the presence of glucose already resulted in values lower than those reported before. Then, the addition of CCCP produced an apparent large decrease of the calculated PMP, because of the release of the large accumulation by the mitochondria and then outside the cells. Interesting also are the lower values obtained in the presence of K^+, whose uptake should lower the PMP.

From previous and recent work, we can summarize the basic conditions needed to obtain reliable results, such as the buffer used, which must not contain cations or organic molecules that may interfere or modify the PMP of the cells. When using different monitors, different conditions should be tested, mainly the concentration of the dye to adjust it depending on the yeast used. Changes due to binding to the surface of the cells can be minimized by the addition of a low concentration (10 μM) $BaCl_2$. The accumulation of the monitor by mitochondria can be avoided by the addition of around 10 μM CCCP or FCCP. Finally, corrections can be applied by using the correct volumes for the distribution of the dye, as well as its efflux with a permeabilizing agent.

There is another factor influencing the results obtained. The efflux of the dye produced by chitosan was lower than that found after the addition of K^+. When the dye concentration after the addition of the monovalent cation is subtracted from that obtained after the addition of chitosan, a negative net accumulation results, meaning that the cells would have a positive PMP value, which is hard to accept. The most probable explanation is that when chitosan is present, the dye goes out of the cells, leaving inside that bound, and the dye binds inside because of its hydrophobic and also its cationic nature. So the addition of K^+ not only reduces the PMP but also produces the liberation of the dye particularly from its internal binding due

	Acridine yellow*	1.0 μM $DiSC_3(3)$**	167 mM ethidium***
Control	−168.4 ± 7.3	−205 ± 6	−225.6 ± 21
CCCP	−105.2 ± 1.8	−169 ± 8	−211.2 ± 31
CCCP+ KCl	−46.0 ± 4.5	−138 ± 8	–

*As described in Ref. [18].
**As described in Ref. [17].
***As described in Ref. [14].

Table 3. Comparison of PMP values (in mV) obtained with different dyes.

to its cationic nature. The conclusion then is that, although the accumulation of acridine yellow provides an adequate method to measure the PMP of yeast, the values obtained after the addition of positively charged ions that accumulate in large concentrations within the cell are distorted because of the displacement of the dye from what most probably are anionic sites inside the cell. It is possible that part of the K^+ taken up by the cells may be bound to their negative internal components. If this were so, one would expect acridine yellow to produce an efflux of K^+. However, in other experiments (unpublished) we have found that the dye at concentrations of 60 and 120 μM, higher than those used in those reported here does not produce the efflux of the monovalent cation. It must also be considered that although the uptake of K^+ is expected to decrease the PMP, the values obtained after the addition of this cation are too low. It has to be considered that after its addition, and because of the decrease of the PMP, this results in the stimulation of the plasma membrane H^+-ATPase, originating a transient increase of ADP, that is then compensated by the acceleration of glycolysis [32], all of which must at least partially restore the PMP values. In this sense, it appears that results following the accumulation of $DiSC_3(3)$ are more in agreement with these facts.

Appendix

In more recent experiments (unpublished), in order to correct as much as possible the values of the binding of the dye to the internal cell components, we incubated the cells under the same conditions, always in the presence of 10 μM CCCP to avoid its accumulation by the mitochondria. Previously, we used 50 μM acridine yellow. When incubation was performed with glucose alone, practically all of the dye was taken up by the cells and made difficult to distinguish between the dye bound inside and that taken up driven by the PMP. Because of this, in these experiments we used the dye at a 150 μM concentration. The conditions were the following:

A. Cells with glucose in which the dye is taken up and accumulated by the cells due partly to the PMP but also to its binding to their internal components.

B. Cells incubated first with glucose for 10 min, adding then 10 mM KCl, in which a large efflux of the dye is observed due in part to the decrease of the PMP, but also to a large K^+ accumulation, around 200–300 mM, that produces its liberation from the anionic sites of the cell. Not considering that this results in values lower than real for the PMP.

C. Cells permeabilized with 100 μg of chitosan for the total 3.0 mL of the incubation mixture. Chitosan liberates the free dye and that bound remains inside.

D. Cells with glucose and with the same concentration of chitosan, but after their permeabilization adding 200 mM KCl. This concentration should displace the dye from the anionic sites to which it supposedly binds because of its cationic nature, but requires a large K^+ concentration to be displaced. Then, the remaining dye inside is that due to its hydrophobicity (**Table 4**).

	Glucose	G + K	Chitosan	Chitosan + K200
[External]	24	85	72	118
nmoles 3.0 mL	71	255	216	355
Entered	379	195	234	95
[Internal]	108,195	55,572	50,554	20,585
[Int]/[Ext]	4551	652		
Log	3.7	2.8		
Apparent Δψ (mV)	**−219**	**−169**		

The experiment was conducted as described in **Table 1**, but to those treated with 100 μg of chitosan, after 10 min, 200 mM KCl was added, and after 5 more min, they were centrifuged. Also, in these experiments, a higher acridine yellow concentration (150 μM) was used. Results from a typical experiment.

Table 4. Accumulation values of acridine yellow in yeast cells and calculations to obtain the apparent values of the PMP.

The experiment was conducted as described for **Table 1**, but to those treated with 100 μg of chitosan, after 10 min, 200 mM KCl was added, and after 5 more min, they were centrifuged. Also, in these experiments, a higher acridine yellow concentration (150 μM) was used, and in all cases, 10 μM CCCP was present. Results from a typical experiment.

The accumulation values using 150 μM acridine yellow concentration were much higher, similar to previous experiments, and values of the PMP without any corrections were −219 and −169 mV, respectively, for the cells incubated only with glucose and with glucose plus 10 mM KCl.

From the values obtained with chitosan, we found that a total concentration of 50,554 nmoles/mL still remained inside the cells, independent from the PMP, presumably bound, both because of the cationic and hydrophobic nature of acridine yellow. The internal concentration, 50,554, was reduced to 20,585 by the addition of 200 mM KCl. This amount remaining in the cells after the addition of chitosan and KCl is that bound due to its hydrophobic nature. The difference (50,554 − 20,585) can be considered that bound because of its cationic nature and is equal to 29,269. Al values are given in nmoles/mL.

These results then, allow the following corrections:

1. With glucose.

A. The internal concentration reached was 108,195 nmoles/mL, but from these, 50,554 were bound independently from the PMP. Subtracting the total bound, the PMP-dependent internal concentration should be 57,641 nmoles/mL. The internal/external concentration ratio would then be 2401.7, its log would be 3.38, and the corrected PMP: −60×3.38 = −201.8 mV.

2. With glucose + KCl.

A. The internal concentration after the addition of KCl decreased to 55,572. However, from these, 50,554 − 20,585, or 29,969 nmoles/mL was displaced by the increased internal concentration of K^+, reached due to its transport, assumed to be near 200 mM. This means that the actual internal concentration of KCl plus that displaced would be 85,541 nmoles/mL. Then, the internal/external concentration ratio would be 1006, and the PMP would change to −180.2.

With these corrections the values obtained in three experiments (means ± std. dev) were −219.5 ± 1.8, with glucose, and −163.3 ± 1.4 with glucose plus 10 mM KCl. The values corrected for the binding due to the cationic nature of the dye were −205.3 ± 3.2 with glucose, and −183.2 ± 3.1 for glucose plus KCl.

As expected, values with glucose, when corrected for the amount of dye bound because of its cationic nature, are somewhat higher than those shown in **Table 1**. The value with glucose plus KCl is much higher, also as expected, because to the amount of dye remaining inside, that displaced by the large accumulation of K^+ was added.

These results confirm that the PMP values obtained are higher than those suggested by other authors. In fact, only with glucose, corrected values are only around 15 mV higher; in the presence of K^+, the values are even higher. This in fact is not unexpected, because, although the addition of K^+, due to its transport mechanism should produce a decrease of the PMP, since our old studies [2, 32] it is known that K^+, by decreasing the PMP, accelerates the plasma membrane H^+-ATPase, which transiently increases the ADP levels, but this increase is rapidly compensated by accelerating glycolysis and respiration. This series of events, but mainly the acceleration of proton pumping, should compensate for the PMP decrease produced by the uptake of K^+. To our knowledge, these measurements with the shown corrections are the most accurate measurements of the PMP in *S. cerevisiae*.

Author details

Antonio Peña*, Norma Silvia Sánchez and Martha Calahorra

*Address all correspondence to: apd@ifc.unam.mx

Department of Molecular Genetics, Instituto de Fisiología Celular, Universidad Nacional Autónoma de México, Mexico City, Mexico

References

[1] Conway E, Brady TG, Carton E. Biological production of acid and alkali. II. A redox theory for the process in yeast with application to the production of gastric acidity. The Biochemical Journal. 1950;**47**:369-374. PMID:14800895

[2] Peña A, Cinco G, Gómez-Puyou A, Tuena M. Effect of the pH of the incubation medium on glycolysis and respiration in *Saccharomyces cerevisiae*. Archives of Biochemistry and Biophysics. 1972;**153**:413-425. DOI: doi.org/10.1016/0003-9861(72)90359-1

[3] Peña A. Studies on the mechanism of K⁺ transport in yeast. Archives of Biochemistry and Biophysics. 1975;**167**:397-409. DOI: doi.org/10.1016/0003-9861(75)90480-4

[4] Gaber RF, Styles CA, Fink GR. TRK1 encodes a plasma membrane protein required for high-affinity potassium transport in *Saccharomyces cerevisiae*. Molecular and Cellular Biology. 1988;**8**:2848-2858. DOI: 10.1128/MCB.8.7.2848

[5] Slayman CW, Long WS, Lu CH. The relationship between ATP and an electrogenic pump in the plasma membrane of *Neurospora crassa*. The Journal of Membrane Biology. 1973;**14**:305-338. DOI: 10.1007/BF01868083

[6] Vacata V, Kotyk A, Sigler K. Membrane potentials in yeast cells measured by direct and indirect methods. Biochimica et Biophysica Acta. 1981;**643**:265-268. DOI: doi.org/10.1016/0005-2736(81)90241-8

[7] Boxman AW, Barts PWJA, Borst-Pauwels GWFH. Some characteristics of tetraphenylphosphonium uptake into *Saccharomyces cerevisiae*. Biochimica et Biophysica Acta. 1982;**686**:13-18. DOI: doi.org/10.1016/0005-2736(82)90146-8

[8] Azzi A, Chance B, Radda GK, Lee CP. A fluorescence probe of energy-dependent structure changes in fragmented membranes. Proceedings of the National Academy of Sciences of the United States of America. 1969;**62**:612-619. PMID: 4307717

[9] Gitler C, Rubalcava B, Caswell A. Fluorescence changes of ethidium bromide on binding to erythrocyte and mitochondrial membranes. Biochimica et Biophysica Acta. 1969;**193**:479-481. DOI: doi.org/10.1016/0005-2736(69)90208-9

[10] Ehrenberg B, Montana V, Wei MD, Wuskell JP, Loew LM. Membrane potential can be determined in individual cells from the nernstian distribution of cationic dyes. Biophysical Journal. 1988;**53**:785-794. DOI: 10.1016/S0006-3495(88)83158-8

[11] Freedman JC, Hoffman JF. The relation between dicarbocyanine dye fluorescence and the membrane potential of human red blood cells set at varying Donnan equilibria. The Journal of General Physiology. 1979;**74**:187-212. DOI: 10.1085/jgp.74.2.187

[12] Sims PJ, Waggoner AS, Wang CH, Hoffman JF. Studies on the mechanism by which cyanine dyes measure membrane potential in red blood cells and phosphatidylcholine vesicles. Biochemistry. 1974;**13**:3315-3330. PMID: 4842277

[13] Peña A, Clemente SM, Borbolla M, Carrasco N, Uribe S. Multiple interactions of ethidium bromide with yeast cells. Archives of Biochemistry and Biophysics. 1980;**201**:420-428. DOI: doi.org/10.1016/0003-9861(80)90530-5

[14] Peña A, Uribe S, Pardo JP, Borbolla M. The use of a cyanine dye in measuring membrane potential in yeast. Archives of Biochemistry and Biophysics. 1984;**231**:217-225. DOI: doi.org/10.1016/0003-9861(84)90381-3

[15] Calahorra M, Sánchez NS, Peña A. Acridine Yellow. A novel use to estimate and measure the plasma membrane potential in *Saccharomyces cerevisiae*. Journal of Bioenergetics and Biomembranes. 2017. DOI: 10.1007/s10863-017-9699-7

[16] Peña A, Sánchez NS, Calahorra M. Estimation of the electric plasma membrane potential difference in yeast with fluorescent dyes: comparative study of methods. Journal of Bioenergetics and Biomembranes. 2010;**42**:419-432. DOI: 10.1007/s10863-010-9311-x

[17] Madrid R, Gómez MJ, Ramos J, Rodríguez-Navarro A. Ectopic potassium uptake in *trk1 trk2* mutants of *Saccharomyces cerevisiae* correlates with a highly hyperpolarized membrane potential. The Journal of Biological Chemistry. 1998;**273**:14838-14844. DOI: 10.1074/jbc.273.24.14838

[18] Gaskova D, Brodska B, Herman P, Vecer J, Malinsky J, Sigler K, Benada O, Plasek J. Fluorescent probing of membrane potential in walled cells: diS-C3(3) assay in *Saccharomyces cerevisiae*. Yeast. 1998;**14**:1189-1197. PMID: 9791890

[19] Plášek J, Gášková D, Lichtenberg-Fraté H, Ludwig J, Höfer M. Monitoring of real changes of plasma membrane potential by diS-C(3)(3) fluorescence in yeast cell suspensions. Journal of Bioenergetics and Biomembranes. 2012;**44**:559-69. DOI: 10.1007/s10863-012-9458-8

[20] Petrezsélyová S, Ramos J, Sychrová H. Trk2 transporter is a relevant player in K^+ supply and plasma-membrane potential control in *Saccharomyces cerevisiae*. Folia Microbiologia (Praha). 2011;56:23-28. DOI: 10.1007/s12223-011-0009-1

[21] Kinclova-Zimmermannova O, Gaskova D, Sychrova H. The Na^+, K^+/H^+antiporter influences the plasma membrane potential of *Saccharomyces cerevisiae*. FEMS Yeast Research. 2006;**6**:792-800. DOI: 10.1111/j.1567-1364.2006.00062.x

[22] Maresova L, Muend S, Zhang Y-Q, Sychrova H, Rao R. Membrane hyperpolarization drives cation influx and fungicidal activity of amiodarone. The Journal of Biological Chemistry. 2009;**284**:2795-2802. DOI: 10.1074/jbc.M806693200

[23] Petrezselyova S, Zahradka J, Sychrova H. *Saccharomyces cerevisiae* BY4741 and W303-1A laboratory strains differ in salt tolerance. Fungal Biology. 2010;**114**:144-150. DOI: doi.org/10.1016/j.funbio.2009.11.002

[24] Montiel V, Ramos J. Intracellular Na and K distribution in *Debaryomyces hansenii*. Cloning and expression in *Saccharomyces cerevisiae* of DhNHX1. FEMS Yeast Research. 2007;**7**:102-109. DOI:10.1111/j.1567-1364.2006.00115.x

[25] Peña A, Sánchez NS, Álvarez H, Calahorra M, Ramírez J. Effects of high medium pH on growth, metabolism and transport in *Saccharomyces cerevisiae*. FEMS Yeast Research. 2015;**15**(2). DOI: 10.1093/femsyr/fou005

[26] Slavik J. Fluorescent Probes in Cellular and Molecular Biology. CRC Press, Inc. Boca Raton, Ann Arbor, London, Tokyo. 1994

[27] Peña A and Ramírez G. Interaction of ethidium bromide with the transport system for monovalent cations in yeast. The Journal of Membrane Biology. 1975;**22**:369-384. DOI: 10.1007/BF01868181

[28] Peña A, Carrasco N, Mora MA. Uptake and effects of several cationic dyes on yeast. The Journal of Membrane Biology. 1979;**47**:261-284. DOI: 10.1007/BF01869081

[29] Kotyk A. Intracellular pH of Baker's yeast. Folia Microbiologia (Praha). 1963;**8**:27-31. DOI: 10.1007/BF02868762

[30] Peña A, Sánchez NS, Calahorra M. Effects of chitosan on *Candida albicans*. Conditions for its antifungal activity. BioMed Research International. 2013;**2013**:1-15. DOI: 10.1155/2013/527549

[31] Sánchez NS, Arreguín R, Calahorra M, Peña A. Effects of salts on aerobic metabolism of *Debaryomyces hansenii*. FEMS Yeast Research. 2008;**8**:1303-1312. DOI:10.1111/j.1567-1364.2008.00426.x

[32] Peña A, Cinco G, Gómez Puyou A, Tuena M. Studies on the mechanism of the stimulation of glycolysis and respiration by potassium ions in *Saccharomyces cerevisiae*. Biochimica et Biophysica Acta. 1969;**180**:1-8. DOI: doi.org/10.1016/0005-2728(69)90187-X

Chapter 2

Metallothioneins, *Saccharomyces cerevisiae*, and Heavy Metals: A Biotechnology Triad?

Farcasanu Ileana Cornelia and Ruta Lavinia Liliana

Additional information is available at the end of the chapter

http://dx.doi.org/10.5772/intechopen.70340

Abstract

Metal ions are the least sophisticated chemical species that interact or bind to biomolecules. The yeast Saccharomyces cerevisiae represents a versatile model organisms used in both basic and applicative research, and one of the main contributors to the understanding of the molecular mechanisms involved in the transport, accumulation, and homeostasis of heavy metals. With a negatively charged wall, the yeast cells are very good biosorbents for heavy metals. In addition to biosorption, the metabolically active cells take up heavy metals via the normal membrane transport systems. Once in the cell, the toxicity of the heavy metals is controlled by various mechanisms, including sequestration by metal-binding proteins, such as the metallothioneins. Metallothioneins are cysteine-rich proteins involved in the buffering of excess heavy metals, both essential (Cu and Zn) and nonessential (Cd, Ag, and Hg). S. cerevisiae has two innate metallothioneins, Cup1 and Crs5, intensively investigated. Additionally, S. cerevisiae served as a host for the heterologous expression of a variety of metallothioneins from different species. This review focuses on the technological implications of expressing metallothioneins in yeast and on the possibility to use these transgenic cells in heavy metal related biotechnologies: bioremediation, recovery of rare metals, or obtaining clonable tags for protein imaging.

Keywords: metallothionein, *Saccharomyces cerevisiae*, heavy metal, bioremediation

1. Introduction

Biotechnology, which makes use of living organisms for technological purposes, is one of the applied fields that constantly benefited from the rapid advancements made in understanding life at molecular level. It is undoubtedly that the budding yeast *Saccharomyces cerevisiae* is one of the biotechnology's most versatile tools. Used since ancient times in bakery, brewery, and wine making, the advent of molecular biology brought new glitter to this eukaryotic microorganism, and *S. cerevisiae* was practically reinvented. With an easy-to-manipulate genomics, with elegant and simple genetics, hosting molecules and biochemical processes well conserved along evolution, this microorganism served as a model organism for the discovery and understanding of numerous essential life mechanisms. Apart from playing an unique and undisputed role in basic research of life science, *S. cerevisiae* kept on adding value to its role in biotechnology, especially due to its capacity to heterologously express proteins, with various scopes: large-scale production of recombinant proteins of technological significance (enzymes, antibodies, and hormones); selection of strains with targeted characteristics and performances (metabolic engineering); environmental biotechnology (yeast surface display); and scientific reasons, such as elucidating the role of proteins from various organisms organism (yeast two-hybrid technique, expression of homologous counterparts from higher organisms in yeast, design yeast as model for human diseases, etc.) [1].

Metallothioneins (MTs) represent one of the numerous examples of proteins whose functions were investigated by heterologous expression in *S. cerevisiae* cells. MTs are low-molecular weight proteins that exist in most organisms from bacteria to humans, including yeasts [2]. MTs constitute an extremely heterogeneous family of cysteine-rich proteins (close to 30% of their amino acid content) that give rise to metal-thiolate complexes ensuing on metal ion coordination. MTs are considered to have many roles, being involved in protection against toxic metals, homeostasis/chaperoning of physiological metals, free-radical scavenging and antioxidative stress protection, control of oxidative state of the cell, anti-apoptotic defense, etc. While some roles still remain obscure, it is widely accepted that all MTs have an undisputed capacity to buffer intracellular metal ions, especially Zn(II) and Cu(I) [3].

Heavy metals belong to a group of nondegradable chemicals naturally present in the environment. Numerous anthropogenic activities, especially the ones related to massive industrialization, intensive agriculture, or rapid urbanization led to important perturbations (accumulation, or in some cases, depletion) in heavy metal balance, with ecological, nutritional, and environmental impacts [4–10]. Some of the heavy metals (Co, Cu, Fe, Mn, Ni, and Zn) are essential for life in trace amounts, playing a pivotal role in the structure of enzymes and other proteins. Other heavy metals (Cd, Sb, Cr, Pb, As, Co, Ag, Se, and Hg) albeit not essential, interfere easily with the metabolism of essential heavy metals, competing for the various physiological transport systems as well as for the biomolecules they bind to. Essential or not, when present in high concentrations, heavy metals are strongly deleterious to living organisms due to nonspecific binding to proteins, often inducing oxidative stress, or disrupting biological membranes. Defense mechanisms against nonphysiological concentrations of heavy metal ions include excretion, compartmentalization in cell organelles, or increased synthesis of metal-buffering molecules, such as the MTs.

S. cerevisiae has been thoroughly investigated, and many mechanism involved in heavy metal transport and homeostasis have been elucidated in this organism [11–16], preparing the grounds for development of techniques used to engineer *S. cerevisiae* cells for increased heavy metal accumulation and improved tolerance. The present review focuses on the studies that relate heterologous expression of MTs in *S. cerevisiae* to the metal-binding characteristics observed and to the possibility to use them for biotechnological purposes.

2. Innate and heterologous expression of MTs in *S. cerevisiae*

Apart from being classified on the basis of their structural homology or on taxonomic criteria, MTs are also classified on molecular functionality grounds, starting from their innate metal-binding abilities, into Cu(I)- and Zn(II)-thioneins, with the representative nonessential models Ag(I) and Cd(II), respectively [3, 17]. This is based on the formation of homometallic MT species when they are produced in metal-enriched media; this classification is not regarded as absolute, since cross-affinity is often noticed for Zn(II)-thioneins binding Cu(I) and vice versa [3].

S. cerevisiae has two structurally different MTs, Cup1, and Crs5. Cup1 has been classified as the strictest Cu(I)-isoform (genuine Cu(I)-thionein) [3]. Cup1 biosynthesis is copper-activated at transcriptional level via the copper-binding transcription factor Ace1/Cup2 [18–21] providing the principal method of cellular removal and sequestering the extremely toxic Cu(I) [21–23]. Although substantially divergent from vertebrate and plant MTs, the arrangement of 12 cysteine residues, which is a hallmark of metal-binding proteins, is partially conserved. In contrast to the MTs from higher eukaryotes, Cup1 is responsible only for Cu(I) and Cd(II) ion tolerance *in vivo* [24, 25], albeit capable of binding other metal ions *in vitro* [26]. This is in contrast to the MTs found in higher eukaryotes, which are typically capable of detoxifying an array of metal ions.

Considered a secondary copper-resistant agent in *S. cerevisiae*, Crs5 is nonhomologous to the paradigmatic Cup1, determining survival under Zn(II) overload in a *CUP1*-null background. Its overexpression prevents the deleterious effects exhibited on the *cup1Δ crs5Δ* double knock-out mutant by exposure to combined Zn(II)/Cu(II), similarly to mouse MT1 Zn-thionein, but not to Cup1. Numerous similar observations denoted that Crs5 has a dual metal-binding behavior, being significantly closer to Zn(II)-thioneins than to Cu(I)-thioneins [23, 27, 28].

Following the discovery and characterization of Cup1, many newly discovered MTs were characterized by heterologous expression in *S. cerevisiae* (**Table 1**).

In plants, the first evidence for the role of MTs in Cu(II) and Cd(II) tolerance was provided by expressing two *Arabidopsis thaliana* MT genes in MT-deficient yeast cells. For example, when expressed in *cup1Δ* knock-out mutant, both *At*MT1 and *At*MT2 complemented the *cup1Δ* mutation by providing a high level of resistance to $CuSO_4$ and moderate resistance to $CdSO_4$ [29]. Later, all four types of plant MTs were checked as metal chelators by expressing *A. thaliana* MT cDNAs (*At*MT1a, *At*MT2a, *At*MT2b, *At*MT3, *At*MT4a, and *At*MT4b) in the Cu(II) and

MT expressed	Source organism	Behavior in *S. cerevisiae*	Reference
*At*MT1a, *At*MT2a, *At*MT2b, *At*MT3, *At*MT4a, and *At*MT4b	*Arabidopsis thaliana* Model plant organisms, metal nonaccumulator	Complement Cu(II) and Cd(II) sensitivity of a *cup1Δ* mutant; *At*MT4a, and *At*MT4b conferred greater Zn(II) tolerance and higher accumulation of Zn(II) than other MTs to the *zrc1Δ cot1Δ* mutant	[29, 30]
*Jc*MT2a	*Jatropha curcas* L. Technical plant: biofuel production	Complements Cu(II) and Cd(II) sensitivity of a *cup1* mutant	[31]
*Ha*MT1, *Ha*MT2, *Ha*MT3, *Ha*MT4	*Helianthus annuus* Technical plant: nutritional oil. Seeds tend to accumulate Cd(II), Pb(II), and Hg(II)	Complement Cu (II) (all), Cd(II) (*Ha*MT4-1), and Zn(II) (*Ha*MT3, *Ha*MT4-1) sensitivity	[32, 33]
*Nc*MT1, *Nc*MT2	*Noccaea (Thlaspi) caerulescens* Cd(II)/Zn(II) hyperaccumulator	*Nc*MT1, and to a lesser extent *Nc*MT2 complement Cu(II), Cd (II), and Zn(II) sensitivity	[35]
*Sv*MT2b, *Sv*MT3	*Silene vulgaris* (Moench) Garcke Cu(II)-hypertolerant plant	Restore Cd(II) and Cu(II) tolerance to yeast sensitive strains	[36, 37]
*Os*MT1-1b	*Oryza sativa*	Confers tolerance to Cd(II), H_2O_2, and ethanol	[38]
*Os*MT1a	*Oryza sativa* L. cv. Iapar 9	Confers tolerance to Zn(II)	[39]
*rg*MT	*Oryza sativa*	Confers vigorous growth under surplus $CuCl_2$, $FeCl_2$, NaCl, $NaHCO_3$, and H_2O_2	[40]
*Put*MT2	*Puccinellia tenuiflora* Alkaline/saline tolerant grass	Tolerance to H_2O_2, NaCl, $NaHCO_3$, Zn(II), Fe(II), Fe(III), Cd(II), Cr(VI), and Ag(I); sensitivity to Mn(II), Co(II), Cu(II), Ni(II)	[41]
*Cv*MT1	*Chloris virgata* Swartz Alkaline tolerant grass	Tolerance to salinity, alkaline conditions, and oxidative stress	[42]
*Th*MT3	*Tamarix hispida* Alkaline/saline tolerant plant	Tolerance to Cd(II), Zn(II), Cu(II), and NaCl stresses; increased accumulation of Cd(II), Zn(II), NaCl, but not of Cu(II)	[43]
*Gint*MT1	*Glomus intraradices* Arbuscular mycorrhizal fungus; confers heavy metal tolerance to exposed plants	Complements Cu(II) and Cd(II) sensitivity of a *cup1* mutant	[44]
*Hc*MT1 and *Hc*MT2	*Hebeloma cylindrosporum* Ectomycorrhizal fungus; confers heavy metal tolerance to exposed plants	Complement Cu(II) and Cd(II) sensitivity of *cup1* and *yap1* mutants	[45]

MT expressed	Source organism	Behavior in *S. cerevisiae*	Reference
*Pi*MT	*Paxillus involutus* Ectomycorrhizal fungus; confers heavy metal tolerance to exposed plants	Complements Cu(II) and Cd(II), but not Zn(II) sensitivity.	[46]
*Dm*MTo, *Dm*MTn	*Drosophila melanogaster*	Complement Cu(II) sensitivity	[47]
sMTIII	*Porcine brain cDNA* Growth inhibitory factor (GIF)	Confers metal resistance to yeast cells	[48]
*Mm*MT2a	*Mus musculus* Canonical Zn(II)-thionein	Complements Zn(II) sensitivity	[33]
*Mm*MT1	*Mus musculus*	Clonable tag for electron microscopy	[82]
*h*MT2, GFP-hMT2	*Homo sapiens*	Increased Cu(II) tolerance and capacity to remove Cu(II) when expressed from yeast *CUP1* promoter	[72]

Table 1. Heterologous expression of MTs in *S. cerevisiae*.

Zn(II)-sensitive yeast mutants, *cup1Δ* and *zrc1Δ cot1Δ*, respectively. All four types of *At*MTs provided similar levels of Cu(II) tolerance and accumulation to the *cup1Δ* mutant, while the type-4 *At*MTs (*At*MT4a and *At*MT4b) conferred greater Zn(II) tolerance and higher Zn(II) accumulation to the *zrc1Δ cot1Δ* mutant [30]. Metal-gained tolerance was also tested in yeast mutants expressing MTs from technical plants. Thus, the Cu(II) and Cd(II) sensitivity of yeast mutants was complemented by expression of MT2a isolated from *Jatropha curcas* L., a technical plant used for biofuel production [31]. In a different study, expression of MTs from *Helianthus annuus* (sunflower) overcame the Cu(II), Zn(II), or Cd(II) sensitivity, depending on the MT type expressed ([32], **Table 1**). Along with high nutritional value and significant oil content, the seeds of *H. annuus* tend to accumulate Cd(II), Pb(II), and Hg(II) [33], and *Ha*MTs are major candidates to be one of the determinants for the high metal accumulation properties of this plant.

Other MTs studied in yeast were isolated from heavy metal hypertolerant or hyperaccumulating plants. Hyperaccumulating plants belong to a small group of species capable of growing on metalliferous soils without developing toxicity symptoms [34]. The MTs from the intensively studied hyperaccumulator *Noccaea* (*Thlaspi*) *caerulescens* were expressed in yeast, and it was revealed that *Nc*MT1, and to a lesser extent *Nc*MT2, complemented the Cu(II), Cd(II), and Zn(II) sensitive phenotypes [35]. The *Silene vulgaris* (Moench) Garcke population with high levels of copper tolerance was shown to owe this hypertolerance to increased transcripts of *SvMT2b* gene; expression of *Sv*MT2b in yeast restored Cd(II) and Cu(II) tolerance in different hypersensitive strains [36]. In a different study, *Sv*MT3, whose gene has been locally duplicated in a tandem arrangement in *S. vulgaris* genome was shown to restore the Cu(II) tolerance along with increased Cu(II) accumulation in a Cu(II)-sensitive yeast mutant, and that both duplicated genes were functional [37].

Expression of plant MTs in *S. cerevisiae* cells sometimes determined other MTs-related phenotypes, besides metal tolerance and accumulation, indicating that heterologous MTs can be fully functional in yeast cells. Thus a heterologous expression in *S. cerevisiae* of *Os*MTI-1b, a MT isoform from *Oryza sativa* (rice), enhanced Cd(II), H_2O_2, and ethanol tolerance [38], while *Os*MT-1a from a Brazilian variety of rice conferred Zn(II) tolerance [39]; *rg*MT from the same species conferred vigorous growth to transgenic yeast cells when exposed to surplus $CuCl_2$, $FeCl_2$, NaCl (salinity), $NaHCO_3$ (alkalinity), or H_2O_2 (exogenous oxidative stress) [40]. Encompassing a wider range of stresses, expression of *Put*MT2 from the saline/alkaline grass *Puccinellia tenuiflora* increased the tolerance of transgenic yeast cells to H_2O_2, NaCl, $NaHCO_3$, and also to a series of metal ions: Zn(II), Fe(II), Fe(III), Cd(II), Cr(VI), and Ag(I), while conferring sensitivity to Mn(II), Co(II), Cu(II), and Ni(II) [41]. Expression of *Cv*MT1 from the alkaline grass *Chloris virgata* Swartz significantly increased the yeast cell tolerance to salinity, alkaline conditions, and oxidative stress [42]. In the same line of studies, *Th*MT3 isolated from the alkaline/saline-resistant plant *Tamarix hispida* conferred the transgenic yeast cells increase tolerance to Cd(II), Zn(II), Cu(II), and NaCl stresses, triggering increased accumulation of Cd(II), Zn(II), NaCl, but not Cu(II) [43].

Often, plants acquire heavy metal tolerance when growing on contaminated sites due to symbiosis with the radicular, arbuscular mycorrhizal fungi that penetrate the cortical cells of the roots of a vascular plant; one MT isolated from such fungus, *Glomus intraradices*, was also shown to complement the Cu(II) and Cd(II) sensitivity of a *cup1* mutant [44], while MT1 and MT2 from the ectomycorrhizal fungi *Hebeloma cylindrosporum* and *Paxillus involutus* functionally complemented the Cu(II) and Cd(II) sensitivity of yeast mutants [45, 46].

Studies on animal MTs expressed in yeast are less numerous [33, 47, 48, 72] and are used mainly for technical purposes. One notable example though is mouse *Mm*MT1a, a canonical Zn(II)-thionein (yeast Cup1 is considered a canonical Cu(I)-thionein) [3] shown to confer tolerance when expressed in Zn(II)-sensitive yeast mutants [31]. *S. cerevisiae* was also used to express human MTs, but mainly as a host for large-scale production of hMTs [49–51], for which the more productive methylotrophic yeast *Pichia pastoris* is currently preferred [52].

3. Biotechnological relevance of MTs expression in *S. cerevisiae*

The main function of MTs resides in their structure: small proteins with a significant number of cysteine residues (15–30% of the total amino acid number) [53], a characteristic that confers them a remarkable capacity to bind heavy metal ions by forming metal-thiolate clusters. MTs are natively bound to Cu(I) or Zn(II), exhibiting various affinities for the two metals, in between the canonical Cu(I)-thionein (*S. cerevisiae* Cup1) and canonical Zn(II)-thionein (*C. elegans* MT1) [3]. Ag(I) and Cd(II) can be used as respective models of Cu(II) and Zn(II) for the study of the metal-binding sites of MTs, particularly in those techniques that require isotopically active nuclei (note that copper is in the cuprous form Cu(I) when bound to MT, but the environment contains the more stable cupric ion Cu(II); when taken up by the cell, Cu(II) can be reduced to Cu(I) by Fe(III)/Cu(II)-reductases, or simply by the reductive milieu of the cell).

With high thermodynamic stability combined with kinetic lability, MTs are important candidates for biotechnology applications. In the nonmetalate form, MTs are highly reactive and can virtually bind to any d^{10} metal [53], a trait that makes them interesting candidates for biotechnology. In this case, two aspects of MT reactivity are highly relevant: (1) metal uptake and release and (2) metal exchange [54]. Due to the polydentate thiolate nature of all MTs and their high affinity for most heavy metal ions, there are data available for binding of Cu(I), Cu(II), Cd(II), Hg(II), Ag(I), Au(I), Bi(III), As(III), Co(II), Fe(II), Pb(II), Pt(II), and Tc(IV) [55]. Another important feature of MT reactivity is the dynamic behavior, with metal uptake and release between species of different degrees of metalation. It is widely accepted that the binding of metal ions to MTs occurs rapidly, between 10 and 30 min, although longer stabilization times are required for certain ions, such as Hg(II) or Pb(II) [3].

Studies on metal exchange in MTs have also been done (usually with either Zn(II)- or Cu(I)-thioneins), starting with a metal-loaded MT forced to exchange its initially bound metal ions with other ions. Considering the series of affinity order of heavy metal ions for the thiolate ligands: Fe(II) ≈ Zn(II) ≈ Co(II) < Pb(II) < Cd(II) < Cu(I) < Au(I) ≈ Ag(I) < Hg(II) < Bi(III) [56], the Zn(II)-loaded MTs would be more reactive than Cu(I)-loaded MTs. It was noted that metal exchange occurs at a much slower pace than metal binding to apo-MTs. For example, it was revealed that binding of four equivalents of Cu(I) to Zn(II)-Cup1 required a stabilization time of 24 h to produce a mixture of Cu4-Cup1 and Cu8-Cup1 species by total displacement of the initially bound Zn(II) [3, 57]. It is interesting to note that many xenobiotic metal ions (Cd(II), Pb(II), and Hg(II)) show higher affinity for thiolate ligands than Zn(II) or Cu(I) does, and thus, in case of intoxication, MTs can work as detoxifying agents [53]. This is highly relevant especially when designing a biotechnology system aimed for removal of toxic ions, as in the case of bioremediation. In the following paragraph, studies on metal accumulation by MTs expressing *S. cerevisiae* cells are presented, and also summarized in **Table 2**.

3.1. Display of MTs on the surface of *S. cerevisiae* cells

Cell surface engineering has wide applicability due to the fact that virtually any protein can be produced and autoimmobilized on the cell exterior of an engineered cell (usually a microorganism). *S. cerevisiae* is suitable for this technique by which functional heterologous proteins/peptides can be displayed on cell surface by fusion with parts of cell wall- or cell membrane-anchoring proteins [58–62]. *S. cerevisiae*, generally regarded as safe (GRAS), is a more suitable host for cell surface engineering than other microorganisms in which the cell surface display system has been established, because yeast possesses a quality-control system for proteins and glycosylation systems of secreted proteins. In addition to the general advantages, high-molecular-mass proteins or proteins that require glycosylation modification can be displayed on yeast cell surface with maintenance of their activities, unlike when displaying them on bacteria [63]. Surface engineered cells can be subsequently treated as microparticles covered with the targeted protein [64].

S. cerevisiae cells are very good biosorbents for heavy metal ions due to the cell wall constituents, which readily sequester heavy metals once they encounter them. These constituents possess numerous metal-loving functional groups, including carboxylate, phosphate, sulfate, and

Metal investigated	Expressed MT	MT provenience	Yeast gained characteristics due to expression of MT	Reference
Cd(II)	Cup1/His6 Yeast surface display	*S. cerevisiae*	Cd(II) tolerance, Cd(II) increased adsorption; selectivity against Cu(II)	[66]
Cd(II)	$^{\Delta1\text{-}8}$Cup1 ($^{\Delta1\text{-}8}$Cup1)$_4$ ($^{\Delta1\text{-}8}$Cup1)$_8$ Surface display of tandem repeats of head-to-tail yeast MT lacking the first 8 amino acids	*S. cerevisiae*	Cd(II) tolerance and adsorption were dependent on the number of tandem repeats; 4 and 8 repeats determined increased Cd(II) adsorption/recovery 5.9 and 8.9 times, respectively	[67]
Cd(II)	*Sn*MT2a, *Sn*MT2c, *Sn*MT2d, and *Sn*MT2e Yeast surface display	*Solanum nigrum* (Cd(II)/Zn(II) hyperaccumulator)	Increased Cd(II) tolerance and adsorption; concentration of Cd(II) from ultra-trace media; selectivity to Cd(II) against Cu(II) and Hg(II)	[69]
Cd(II)	*Sa*MT2	*Sedum alfredii* Hance (Cd/Zn hyperaccumulator)	Increased Cd(II) tolerance and accumulation	[76]
Cd(II)	*Th*MT3	*Tamarix hispida* Alkaline/saline tolerant plant	Increased Cd(II) tolerance and accumulation	[43]
Cu(II)	hMT2, GFP-hMT2	*Homo sapiens*	Increased Cu(II) tolerance and capacity to remove Cu(II) when expressed from yeast *CUP1* promoter	[72]
Zn(II)	*At*MT4a and *At*MT4b	*Arabidopsis thaliana*	Increased accumulation of Zn(II)	[30]
Zn(II)	*Th*MT3	*Tamarix hispida* Alkaline/saline tolerant plant	Increased Zn(II) tolerance and accumulation	[43]

Table 2. MTs heterologously expressed in yeast that determine increased accumulation of metal ions.

sulfhydryl, which decorate the outer mannan-protein layer of the wall [65]. The metal-binding innate capacity of the cell wall can be substantially increased by expressing metal-binding peptides/proteins at the cell surface [59, 61]. Using this technique, yeast cells were modified for bioremediation of Cd(II) using a cell-surface display system of its own MT, Cup1, fused with a hexahistidyl residue, by using an α-agglutinin-based display system [66]. Surface-engineered yeast cells with Cup1 and hexa-His fused in tandem (Cup1-His6, originally named YMT-hexaHis) showed superior cell-surface adsorption and recovery of Cd(II) under EDTA treatment on the cell surface compared to the His6-displaying cells, through an additive effect on chelating ability. Remarkably, the expression of Cup1-His6 did not have a strong effect on the adsorption of Cu(II). The same study revealed that yeast cells displaying Cup1-His6 exhibited a higher potential for the adsorption of Cd(II) than *Escherichia coli* cells displaying the same constructs. Additionally, cells displaying tandem Cup1-His6 showed increased resistance to Cd(II) through active and enhanced adsorption of the toxic ion, indicating that Cup1-His6-displaying yeast cells are unique biosorbents with a superior functional chelating ability.

Adsorption of heavy metal ions at the cell surface has certain advantages compared to intracellular accumulation. First, surface adsorption allows recycling of the adsorbed ions, whereas intracellular accumulation necessitates disintegration of the cell for extraction. Second, surface adsorption is possible even in nonviable cells, providing that sufficient biomass can be produced. This is particularly important when cells are used to remove heavy metals from contaminated waters, and the conditions necessary to sustain living cells are difficult to achieve. And third, surface-engineered yeast cells can be used repeatedly as bioadsorbents since the recovery and treatment of the heavy metal ions does not greatly damage the cells [66]. In a sequel study, Cup1 was expressed as tandem head-to-tail repeats of the yeast MT lacking the first 8 amino acids (known to be nonsignificant for metal binding). Three types of constructs that were surface displayed contained 1, 4, and 8 tandem MT repeats [67].

The transgenic cells obtained were tested against excess Cd(II), and it was revealed that the adsorption and recovery of Cd(II) on the cell surface was increasingly enhanced with increasing the number of tandem repeats under conditions that allowed complete occupation of the Cd(II)-binding sites in the MT tandem repeats. Considering the relationship between cell-surface adsorption and protection against heavy metal ion toxicity, the tolerance of these surface-engineered yeasts to Cd(II) was found to be also dependent on the number of displayed MT tandem repeats, indicating that the characteristics of surface-engineered yeasts as a bioadsorbents correlated with the ability of the displayed proteins to bind metal ions [67]. Unfortunately, these promising studies soon came to a halt and no other metal ions or other MTs were taken into consideration to be used in this technique. It took ten years before another group displayed at the surface of yeast cells four type-2 MTs from *Solanum nigrum* (*Sn*MT): *Sn*MT2a, *Sn*MT2c, *Sn*MT2d, and *Sn*MT2e [68]. *S. nigrum* is an ornamental shrub (nightshade) and a Cd(II)/Zn(II) hyperaccumulator, apparently due to the four *Sn*MTs subtypes (*Sn*MT2a, *Sn*MT2c, *Sn*MT2d, and *Sn*MT2e) shown to have an important role in metal detoxification [69]. Yeast strains displaying the *Sn*MTs specified above on the cell surface were obtained, and these strains were shown to develop both Cd(II) tolerance and increased Cd(II) adsorption, exhibiting a higher affinity for Cd(II) than for Cu(II) or Hg(II) [68]. Notably, these displaying strains could effectively adsorb ultra-trace Cd(II) and accumulate it under a wide range of pHs (between 3 and 7), without disturbing the co-exising Cu(II) and Hg(II) [68]. Moreover, apart from showing a high potential for removing Cd(II) from contaminated waters, the yeast-surface engineered strains expressing *Sn*MT showed a remarkable resistance to Cd(II): while the nonengineered cells were stopped from dividing by 80 μM Cd(II), the engineered strains could live in 500 μM Cd(II) [68], a very high concentration for aqueous environments. While the study does not present accumulation data on other heavy metal ions, it is notable that the engineered strains expressing *Sn*MT could concentrate ultra-trace Cd(II) on the cell surface, encouraging further attempts to display other MTs on yeast surface (from hyperaccumulating species) with the final scope of concentrating rare metal ions from ultra-traces environments.

3.2. MT-expressing *S. cerevisiae* cells for bioremediation

Heavy metal bioremediation is an appealing approach for decontaminating polluted environments, especially because standard physico-chemical methods are ineffective and very often a source of pollution themselves [5]. An ideal heavy metal bioremediator would have certain metal-related characteristics: tolerance to high concentrations, increased accumulation, and substantial biomass production for effective removal of heavy metal ions from the contaminated sites. These traits fall into the characteristics of the heavy metal hyperaccumulating plants, with the exception that they usually do not produce sufficient biomass [70]. *S. cerevisiae* is an example of an organism that could be engineered for bioremediation purposes. The cell surface display of metal-binding peptides/proteins presented above may not be the best approach for obtaining hyperaccumulating yeasts, since the metal binding is restricted to cell surface. Rather, (over)expressing nontoxic metal-binding proteins within yeast cell may increase the chances of obtaining hyperaccumulating strains fit for bioremediation purposes. *S. cerevisiae* is not a heavy metal accumulator due to a very active excretion system which extrudes excess metal ions from the cell [71, 72]. However, the excess free ions could be retained within the cell in a nontoxic form through sequestration by an abundant metal-binding protein, such as an overexpressed MT. Recently, an increased Cu(II) bioremediation ability of new transgenic and adapted *S. cerevisiae* strains was described [73]. In this study, *S. cerevisiae* cells were manipulated to integrate human MT2 (hMT2) and GFP-hMT2, expressed from either the constitutively p*ADH1* yeast promoter or the Cu(II)-inducible p*CUP1* yeast promoter. It was shown that only cells that expressed hMT2 from the *CUP1* promoter exhibited both increased Cu(II) tolerance and capacity to remove Cu(II) ions from growth media [73].

3.3. Heterologous expression of MTs from heavy metal hyperaccumulators

The natural heavy metal hyperaccumulators, mostly belonging to a small group of plants [34, 70], are the species whose metal-related characteristics initially prompted the ideas of bioremediation, biomining, and bioextraction. To accumulate heavy metals without developing toxicity symptoms, these organisms utilize a variety of chemical ligands capable of coordinating the metal ions in a nontoxic form. Although MTs are important candidates for sequestering heavy metal ions, the studies relevant for correlating MT expression with heavy metal accumulating phenotype are scarce and hardly encouraging [74, 75]. The examples of MT from hyperaccumulating organisms expressed in yeast are few, and they mainly focus on functional complementation tests [33, 36–38, 69, 76]. One example is worth mentioning here though, as it deals with an unusual hyperaccumulating phenotype: Ag(I)-hyperaccumulation due to three distinct MT genes of the ectomycorrhizal fungus *Amanita strobiliformis* [77, 78]. Although expressed in *S. cerevisiae* only to test the restoration of Cd(II), Cu(II), and Zn(II) tolerance to sensitive mutants, further employment of *As*MT for cellular handling of Ag(I) is worth considering.

3.4. Metallothionein as clonable tags

Due to their small size and metal-binding capacity, metallothioneins may be interesting candidates for tagging proteins for imaging, especially by electron microscopy (EM) [79–81]. Localization of proteins in cells or complexes using EM relies upon the use of heavy metal clusters, which can be difficult to direct to sites of interest. For this reason, a metal-binding clonable tag, such as it is green fluorescent protein (GFP) for light microscopy, has been pursued for a long time, and would be unvaluable for imaging by EM techniques. In this respect, MT is a very good candidate, because instead of fluorescing like GFP, it would initiate formation of a heavy metal cluster adjacent to the protein to be analyzed. A suitable clonable tag for EM is expected to have certain properties: small size and low molecular weight, so as not to disrupt protein kinetics/function *in vivo*. Using MTs as clonable tag implies working on either macromolecular assemblies or cells, but avoiding issues of heavy metal toxicity by delaying the addition of metal until the samples that include a protein-MTH chimera are in preparation for EM. Two successful procedures: (1) adding heavy metal to sections of samples that have already been rapidly frozen, fixed by freeze substitution, and embedded in a hydrophilic plastic and (2) adding metal during the process of freeze substitution have been described [82]. Using *S cerevisiae* as an expression system, it was shown that MT can be localized in the complex environment of a cell, and with a very good signal-to-noise ratio [83]. Thus, mouse MT1 used to tag the yeast centrosomal protein Spc42 allowed the localization the MT-tagged Spc42 in the outer layer of the central plaque of the mature yeast spindle pole body. Nevertheless, although very promising, MT tagging for protein localization may not be universally applicable as this approach did not work with protein components of the nuclear pore complex [82]. Another potential use of MT as clonable tag for imaging would be the yellow luminescence observed for Cu(I)-MT complexes [84, 85].

4. Concluding remarks

The numerous studies on MTs stand for the uniqueness of these small proteins whose undisputed trait is binding to heavy metal ions. This is evidently due to the cysteinyl residues, which represent more than 20% of the total number of MT amino acids, whereas the usual percentage of cysteinyl residues seldom surpasses 5% in most proteins. The intrinsic characteristic of sequestering metal ions in thiolate clusters make MTs very interesting biomolecules for various biotechnological application. Since *S. cerevisiae* represents a very good cellular system for heterologous expression of MTs from virtually any species (including itself), the use of MT-(over)expressing yeasts is a promising starting point for biotechniques such as heavy metal bioremediation and bioextraction. The data summarized in **Table 1** reveal that until now MT-expressing *S. cerevisiae* cells have been used for functional complementation studies (mainly MTs from plants and mycorrhizal fungi) rather than to investigate their biotechnological potential. Moreover, most of the studies concern the ions that naturally bind to MTs, Cu(I) and Zn(II), and their nonessential counterparts Ag(I) and Cd(II), along with few studies on sulfur-loving metal ions such as Hg(II) and Pb(II). Although other interesting noncanonical metal ions such as Mn(II), Ni(II), and Co(II) have been shown to strongly bind to MTs [55, 56], very few studies actually determined the MT binding to these ions *in vivo* (**Table 2**), and MT-expressing yeast cells would be very good models for filling this gap. Especially, obtaining heavy metal hyperaccumulating yeast cells by heterologous (over)expression of recombinant MTs would open new opportunities for bioremediation, bioextraction, and for emerging techniques, such as synthesis of clonable heavy metal nanoparticles [86]. Another promising biotechnique involving MTs is obtaining new clonable tags for cell imaging. While some timid progress has been reported on imaging by EM of proteins tagged with MT in yeast cells [82], the possibility to use the yellow luminescence of Cu(I)-MT for imaging MT-tagged proteins is largely unexplored. In this direction, construction of a systematic collection of *S. cerevisiae* strains that express *all* the MTs identified so far would be not only a challenge but also a prerequisite for systematic investigation of MTs for various biotechnology purposes.

Acknowledgements

The authors received funding from the Romanian—EEA Research Programme operated by the Ministry of National Education under the EEA Financial Mechanism 2009–2014 and Project Contract No 21 SEE/30.06.2014.

Author details

Farcasanu Ileana Cornelia* and Ruta Lavinia Liliana

*Address all correspondence to: ileana.farcasanu@chimie.unibuc.ro

Department of Organic Chemistry, Biochemistry and Catalysis, Faculty of Chemistry, Bucharest University, Romania

References

[1] Feldmann H. Yeasts in biotechnology. In: Feldmann H, editor. Yeast: Molecular and Cell Biology. 2nd ed. Wiley-Blackwell, Weinheim, Germany; 2012. pp. 347-371. ISBN: 978-3-527-64486-5

[2] Capdevila M, Atrian S. Metallothionein protein evolution: A mini assay. Journal of Biological Inorganic Chemistry. 2011;**16**:977-989. DOI: 10.1007/s00775-011-0798-3

[3] Palacios O, Atrian S, Capdevila M. Zn- and Cu-thioneins: A functional classification. Journal of Biological Inorganic Chemistry. 2011;**16**:991-1009. DOI: 10.1007/s00775-011-0827-2

[4] Prasad MNV, editor. Heavy Metal Stress in Plants. From Biomolecules to Ecosystems. Berlin, Heidelberg: Springer Berlin Heidelberg; 2004. p. 454. DOI: 10.1007/978-3-662-07743-6

[5] Bradl H. Heavy Metals in the Environment: Origin, Interaction and Remediation. Vol. 6. 1st ed. London: Academic Press; 2005. p. 282. ISBN: 9780120883813

[6] He ZL, Yang XE, Stoffella PJ. Trace elements in agroecosystems and impacts on the environment. Journal of Trace Elements in Medicine and Biology. 2005;**19**:125-140. DOI: 10.1016/j.temb.2005.02.010

[7] Kim HS, Kim YJ, Seo YR. An overview of carcinogenic heavy metal: Molecular toxicity mechanism and prevention. Journal of Cancer Prevention. 2015;**20**:232-240. DOI: 10.15430/JCP.2015.20.4.232

[8] Jan AT, Azam M, Siddiqui K, Ali A, Choi I, Haq QM. Heavy metals and human health: Mechanistic insight into toxicity and counter defense system of antioxidants. International Journal of Molecular Sciences. 2015;**16**:29592-29630. DOI: 10.3390/ijms161226183

[9] Caito S, Aschner M. Neurotoxicity of metals. Handbook of Clinical Neurology. 2015;**131**:169-189. DOI: 10.1016/B978-0-444-62627-1.00011-1

[10] Wu X, Cobbina SJ, Mao G, Xu H, Zhang Z, Yang L. A review of toxicity and mechanisms of individual and mixtures of heavy metals in the environment. Environmental Science and Pollution Research International. 2016;**23**:8244-8259. DOI: 10.1007/s11356-016-6333-x

[11] Van Ho A, Ward DM, Kaplan J. Transition metal transport in yeast. Annual Review of Microbiology. 2002;**56**:237-261. DOI: 10.1146/annurev.micro.56.012302.160847

[12] Ratherford JC, Bird AJ. Metal-responsive transcription factors that regulate iron, zinc, and copper homeostasis in eukaryotic cells. Eukaryotic Cell. 2004;**3**:1-13. DOI: 10.1128/EC.3.1.1-13.2004

[13] Massé E, Arguin M. Ironing out the problem: New mechanisms of iron homeostasis. Trends in Biochemical Sciences. 2005;**30**:462-468. DOI: 10.1016/jtibs.2005.06.005

[14] Eide DJ. Homeostatic and adaptive responses to zinc deficiency *in Saccharomyces cerevisiae*. The Journal of Biological Chemistry. 2009;**284**:18565-18569. DOI: 10.1074/jbc.R900014200

[15] Reddi AR, Jensen LT, Culotta VC. Manganese homeostasis in *Saccharomyces cerevisiae*. Chemical Reviews. 2009;**109**:4722-4732. DOI: 10.1021/cr900031u

[16] Reddi AR, Leung E, Aranda K, Jensen LT, Culotta VC. The effect of phosphate accumulation on metal ion homeostasis in *Saccharomyces cerevisiae*. Journal of Biological Inorganic Chemistry. 2010;**15**:1051-1062. DOI: 10.1007/s00775-010-0664-8

[17] Valls M, Bofill R, Gonzalez-Duarte R, Gonzalez-Duarte P, Capdevila M, Atrian. A new insight into metallothionein (MT) classification and evolution. The in vivo and in vitro metal binding features of *Homarus americanus* recombinant MT. The Journal of Biological Chemistry. 2001;**276**:32835-32843. DOI: 10.1074/jbc.M102151200

[18] Thiele DJ. *ACE1* regulates expression of the *Saccharomyces cerevisiae* metallothionein gene. Molecular and Cellular Biology. 1988;**8**:2745-2752. DOI: 10.1128/MCB.8.7.2745

[19] Welch J, Fogel S, Buchman C, Karin M. The *CUP2* gene product regulates the expression of the *CUP1* gene, coding for yeast metallothionein. The EMBO Journal. 1989;**8**:255-260

[20] Buchman C, Skroch P, Welch J, Fogel S, Karin M. The *CUP2* gene product, regulator of yeast metallothionein expression, is a copper-activated DNA-binding protein. Molecular and Cellular Biology. 1989;**9**:4091-4095. DOI: 10.1128/MCB.9.9.4091

[21] Butt TR, Sternberg E, Herd J, Crooke ST. Cloning and expression of a yeast copper metallothionein gene. Gene. 1984;**27**:23-33. DOI: 10.1016/0378-1119(84)90235-X

[22] Butt TR, Sternberg EJ, Gorman JA, Clark P, Hamer D, Rosenberg M, Crooke ST. Copper metallothionein of yeast, structure of the gene, and regulation of expression. Proceedings of the National Academy of Sciences of the United States of America. 1984;**81**:3332-3336

[23] Jensen LT, Howard WR, Strain JJ, Winge DR, Culotta VC. Enhanced effectiveness of copper ion buffering by *CUP1* metallothionein compared with *CRS5* metallothionein in *Saccharomyces cerevisiae*. The Journal of Biological Chemistry. 1996;**271**:18514-18519. DOI: 10.1074/jbc.271.31.18514

[24] Ecker DJ, Butt TR, Sternberg EJ, Neeper MP, Debouck C, Gorman JA, Crooke ST. Yeast metallothionein function in metal ion detoxification. The Journal of Biological Chemistry. 1986;**261**:16895-168900

[25] Jeyaprakash A, Welch JW, Fogel S. Multicopy *CUP1* plasmids enhance cadmium and copper resistance levels in yeast. Molecular & General Genetics. 1991;**225**:363-368

[26] Winge DR, Nielson KB, Gray WR, Hamer DH. Yeast metallothionein. Sequence and metal-binding properties. The Journal of Biological Chemistry. 1985;**260**:14464-14470

[27] Culotta VC, Howard WR, Liu XF. CRS5 encodes a metallothionein-like protein in *Saccharomyces cerevisiae*. The Journal of Biological Chemistry. 1994;**269**:25295-25302

[28] Pagani A, Villarreal L, Capdevila M, Atrian S. The *Saccharomyces cerevisiae* Crs5 Metallothionein metal-binding abilities and its role in the response to zinc overload. Molecular Microbiology. 2007;**63**:256-269. DOI: 10.1111/j.1365-2958.2006.05510.x

[29] Zhou J, Goldsbrough PB. Functional homologs of fungal metallothionein genes from *Arabidopsis*. Plant Cell. 1994;**6**:875-884. DOI: 10.1105/tpc.6.6.875

[30] Guo WJ, Meetam M, Goldsbrough PB. Examining the specific contributions of individual *Arabidopsis* metallothioneins to copper distribution and metal tolerance. Plant Physiology. 2008;**146**:1697-1706. DOI: 10.1104/pp.108.115782

[31] Mudalkar S, Golla R, Sengupta D, Ghatty S, Reddy AR. Molecular cloning and characterisation of metallothionein type 2a gene from *Jatropha curcas* L., a promising biofuel plant. Molecular Biology Reports. 2014;**41**:113-124. DOI: 10.1007/s11033-013-2843-5

[32] Tomas M, Pagani MA, Andreo CS, Capdevila M, Atrian S, Bofill R. Sunflower metallothionein family characterisation. Study of the Zn(II)- and Cd(II)-binding abilities of the HaMT1 and HaMT2 isoforms. Journal of Inorganic Biochemistry. 2015;**48**:35-48. DOI: 10.1016/j.jinorgbio.2015.02.016

[33] Madejón P, Murillo JM, Marañón T, Cabrera F, Soriano MA. Trace element and nutrient accumulation in sunflower plants two years after the Aznalcóllar mine spill. The Science of the Total Environment. 2003;**307**:239-257. DOI: 10.1016/S0048-9697(02)00609-5

[34] Krämer U. Metal hyperaccumulation in plants. Annual Review of Plant Biology. 2010;**61**:517-534. DOI: 10.1146/annurev-arplant-042809-112156

[35] Roosens NH, Leplae R, Bernard C, Verbruggen N. Variations in plant metallothioneins: The heavy metal hyperaccumulator *Thlaspi caerulescens* as a study case. Planta. 2005;**222**:716-729. DOI: 10.1007/s00425-005-0006-1

[36] van Hoof NA, Hassinen VH, Hakvoort HW, Ballintijn KF, Schat H, Verkleij JA, Ernst WH, Karenlampi SO, Tervahauta AI. Enhanced copper tolerance in *Silene vulgaris* (Moench)

Garcke populations from copper mines is associated with increased transcript levels of a 2b-type metallothionein gene. Plant Physiology. 2001;**126**:1519-1526. DOI: 10.1104/pp.126.4.1519

[37] Nevrtalova E, Baloun J, Hudzieczek V, Cegan R, Vyskot B, Dolezel J, Safar J, Milde D, Hobza R. Expression response of duplicated metallothionein 3 gene to copper stress in *Silene vulgaris* ecotypes. Protoplasma. 2014;**251**:1427-1439. DOI: 10.1007/s00709-014-0644-x

[38] Ansarypour Z, Shahpiri A. Heterologous expression of a rice metallothionein isoform (OsMTI-1b) in *Saccharomyces cerevisiae* enhances cadmium, hydrogen peroxide and ethanol tolerance. Brazilian Journal of Microbiology. 2017;**48**(16):30354-30359. pii: S1517-8382. DOI: 10.1016/j.bjm.2016.10.024

[39] Yang Z, Wu Y, Li Y, Ling HQ, Chu C. OsMT1a, a type 1 metallothionein, plays the pivotal role in zinc homeostasis and drought tolerance in rice. Plant Molecular Biology. 2009;**70**:219-229. DOI: 10.1007/s11103-009-9466-1

[40] Jin S, Sun D, Wang J, Li Y, Wang X, Liu S. Expression of the rgMT gene, encoding for a rice metallothionein-like protein in *Saccharomyces cerevisiae* and *Arabidopsis thaliana*. Journal of Genetics. 2014;**93**:709-718

[41] Zhang M, Takano T, Liu S, Zhang X. Abiotic stress response in yeast and metal-binding ability of a type 2 metallothionein-like protein (PutMT2) from *Puccinellia tenuiflora*. Molecular Biology Reports. 2014;**41**:5839-5849. DOI: 10.1007/s11033-014-3458-1

[42] Nishiuchi S, Liu S, Takano T. Isolation and characterization of a metallothionein-1 protein in *Chloris virgata* Swartz that enhances stress tolerances to oxidative, salinity and carbonate stress in *Saccharomyces cerevisiae*. Biotechnology Letters. 2007;**29**:1301-1305. DOI: 10.1007/s10529-007-9396-4

[43] Yang J, Wang Y, Liu G, Yang C, Li C. *Tamarix hispida* metallothionein-like ThMT3, a reactive oxygen species scavenger, increases tolerance against Cd(2+), Zn(2+), Cu(2+), and NaCl in transgenic yeast. Molecular Biology Reports. 2011;**38**:1567-1574. DOI: 10.1007/s11033-010-0265-1

[44] González-Guerrero M, Cano C, Azcón-Aguilar C, Ferrol N. GintMT1 encodes a functional metallothionein in *Glomus intraradices* that responds to oxidative stress. Mycorrhiza. 2007;**17**:327-335. DOI: 10.1007/s00572-007-0108-7

[45] Ramesh G, Podila GK, Gay G, Marmeisse R, Reddy MS. Different patterns of regulation for the copper and cadmium metallothioneins of the ectomycorrhizal fungus *Hebeloma cylindrosporum*. Applied and Environmental Microbiology. 2009;**75**:2266-2274. DOI: 10.1128/AEM.02142-08

[46] Bellion M, Courbot M, Jacob C, Guinet F, Blaudez D, Chalot M. Metal induction of a *Paxillus involutus* metallothionein and its heterologous expression in *Hebeloma cylindrosporum*. The New Phytologist. 2007;**174**:151-158. DOI: 10.1111/j.1469-8137.2007.01973.x

[47] Silar P, Wegnez M. Expression of the *Drosophila melanogaster* metallothionein genes in yeast. FEBS Letters. 1990;**269**:273-278. DOI: 10.1016/0014-5793(90)81172-K

[48] Wang SH, Chang CY, Chen CF, Tam MF, Shih YH, Lin LY. Cloning of porcine neuron growth inhibitory factor (metallothionein III) cDNA and expression of the gene in *Saccharomyces cerevisiae*. Gene. 1997;**203**:189-197. DOI: 10.1016/S0378-1119(97)00513-1

[49] Cismowski MJ, Huang PC. Purification of mammalian metallothionein from recombinant systems. Methods in Enzymology. 1991;**205**:312-319

[50] Cismowski MJ, Huang PC. Effect of cysteine replacements at positions 13 and 50 on metallothionein structure. Biochemistry. 1991;**30**:6626-6632

[51] Cismowski MJ, Narula SS, Armitage IM, Chernaik ML, Huang PC. Mutation of invariant cysteines of mammalian metallothionein alters metal binding capacity, cadmium resistance, and 113Cd NMR spectrum. The Journal of Biological Chemistry. 1991;**266**:24390-24397

[52] Wu Q, Li B, Wu F, Yang L, Li S, Li H, Wu D, Cui T, Tang D. High-level expression, efficient purification, and bioactivity of recombinant human metallothionein 3 (rhMT3) from methylotrophic yeast *Pichia pastoris*. Protein Expression and Purification. 2014;**101**:121-126. DOI: 10.1016/j.pep.2014.06.009

[53] Capdevilaa M, Bofilla R, Palaciosa O, Atrianb S. State-of-the-art of metallothioneins at the beginning of the 21st century. Coordination Chemistry Reviews. 2012;**256**:46-62. DOI: 10.1016/j.ccr.2011.07.006

[54] Blindauer CA, Leszczyszyn OI. Metallothioneins: Unparalleled diversity in structures and functions for metal ion homeostasis and more. Natural Product Reports. 2010;27:720-741. DOI: 10.1039/b906685n

[55] Bell SG, Vallee BL. The metallothionein/thionein system: An oxidoreductive metabolic zinc link. Chembiochem. 2009;**10**:55-62. DOI: 10.1002/cbic.200800511

[56] Vasak M. Metal removal and substitution in vertebrate and invertebrate metallothioneins. In: Riordan JF, Vallee BL, editors. Metallobiochemistry Part B, Metallothionein and Related Molecules. Vol. 205. 1st ed. San Diego: Academic Press; 1991. pp. 452-458. ISBN: 9780121821067

[57] Orihuela R, Monteiro F, Pagani A, Capdevila M, Atrian S. Evidence of native metal-S(2-)-metallothionein complexes confirmed by the analysis of Cup1 divalent-metal-ion binding properties. Chemistry. 2010;**16**:12363-12372. DOI: 10.1002/chem.201001125

[58] Shibasaki S, Ueda M. Bioadsorption strategies with yeast molecular display technology. Biocontrol Science. 2014;**19**:157-164. DOI: 10.4265/bio.19.157

[59] Li PS, Tao HC. Cell surface engineering of microorganisms towards adsorption of heavy metals. Critical Reviews in Microbiology. 2015;**41**:140-149. DOI: 10.3109/1040841X.2013.813898

[60] Cherf GM, Cochran JR. Applications of yeast surface display for protein engineering. Methods in Molecular Biology. 2015;**1319**:155-175. DOI: 10.1007/978-1-4939-2748-7_8

[61] Liu Z, Ho SH, Hasunuma T, Chang JS, Ren NQ, Kondo A. Recent advances in yeast cell-surface display technologies for waste biorefineries. Bioresource Technology. 2016;**215**:324-333. DOI: 10.1016/j.biortech.2016.03.132

[62] Ueda M. Establishment of cell surface engineering and its development. Bioscience, Biotechnology, and Biochemistry. 2016;**80**:1243-1253. DOI: 10.1080/09168451.2016.1153953

[63] Kuroda K, Ueda M. Generation of arming yeasts with active proteins. In: Mapelli V, editor. Yeast Metabolic Engineering: Methods and Protocols. Methods in Molecular Biology, 1st ed. Humana Press, Springer Science + Business Media, Springer New York Heidelberg Dordrecht London; 2014. pp. 137-157. DOI: 10.1007/978-1-4939-0563-8_8

[64] Kuroda K, Ueda M. Engineering of microorganisms towards recovery of rare metals. Applied Microbiology and Biotechnology. 2010;**87**:53-60. DOI: 10.1007/s00253-010-2581-8

[65] Brady D, Stoll AD, Starke L, Duncan JR. Chemical and enzymatic extraction of heavy metal binding polymers from isolated cell walls of *Saccharomyces cerevisiae*. Biotechnology and Bioengineering. 1994;**44**:297-302. DOI: 10.1002/bit.260440307

[66] Kuroda K, Ueda M. Bioadsorption of cadmium ion by cell surface-engineered yeasts displaying metallothionein and hexa-His. Applied Microbiology and Biotechnology. 2003;**63**:182-186. DOI: 10.1007/s00253-003-1399-z

[67] Kuroda K, Ueda M. Effective display of metallothionein tandem repeats on the bioadsorption of cadmium ion. Applied Microbiology and Biotechnology. 2006;**70**:458-463. DOI: 10.1007/s00253-005-0093-8

[68] Wei Q, Zhang H, Guo D, Ma S. Cell surface display of four types of *Solanum nigrum* metallothionein on *Saccharomyces cerevisiae* for biosorption of cadmium. Journal of Microbiology and Biotechnology. 2016;**26**:846-853. DOI: 10.4014/jmb.1512.12041

[69] Ferraz P, Fidalgo F, Almeid A, Teixeira J. Phytostabilization of nickel by the zinc and cadmium hyperaccumulator *Solanum nigrum* L. Are metallothioneins involved?. Plant Physiology and Biochemistry. 2012;**57**:254-260. DOI: 10.1016/j.plaphy.2012.05.025

[70] Krämer U. Phytoremediation: Novel approaches to cleaning up polluted soils. Current Opinion in Biotechnology. 2005;**16**:133-141. DOI: 10.1016/j.copbio.2005.02.006

[71] Dürr G, Strayle J, Plemper R, Elbs S, Klee SK, Catty P, Wolf DH, Rudolph HK. The medial-Golgi ion pump Pmr1 supplies the yeast secretory pathway with Ca2+ and Mn2+ required for glycosylation, sorting, and endoplasmic reticulum-associated protein degradation. Molecular Biology of the Cell. 1998;**9**:1149-1162. DOI: 10.1091/mbc.9.5.1149

[72] Lauer Júnior CM, Bonatto D, Mielniczki-Pereira AA, Schuch AZ, Dias JF, Yoneama ML, Pegas Henriques JA. The Pmr1 protein, the major yeast Ca^{2+}-ATPase in the Golgi, regulates intracellular levels of the cadmium ion. FEMS Microbiology Letters. **285**:79-88. DOI: 10.1111/j.1574-6968.2008.01214.x

[73] Geva P, Kahta R, Nakonechny F, Aronov S, Nisnevitch M. Increased copper bioremediation ability of new transgenic and adapted *Saccharomyces cerevisiae* strains.

Environmental Science and Pollution Research International. 2016;**23**:19613-19625. DOI: 10.1007/s11356-016-7157-4

[74] Küpper H, Götz B, Mijovilovich A, Küpper FC, Meyer-Klaucke W. Complexation and toxicity of copper in higher plants. I. Characterization of copper accumulation, speciation, and toxicity in *Crassula helmsii* as a new copper accumulator. Plant Physiology. 2009;**151**:702-714. DOI: 10.1104/pp.109.139717

[75] Mijovilovich A, Leitenmaier B, Meyer-Klaucke W, Kroneck PMH, Goötz B, Küpper H. Complexation and toxicity of copper in higher plants. II. Different mechanisms for copper versus cadmium detoxification in the copper-sensitive cadmium/zinc hyperaccumulator *Thlaspi caerulescens* (Ganges ecotype). Plant Physiology. 2009;**151**:715-731. DOI: 10.1104/pp.109.144675

[76] Zhang J, Zhang M, Tian S, Lu L, Shohag MJ, Yang X. Metallothionein 2 (SaMT2) from *Sedum alfredii* Hance confers increased Cd tolerance and accumulation in yeast and tobacco. PloS One. 2014;**9**:e102750. DOI: 10.1371/journal.pone.0102750

[77] Beneš V, Hložková K, Matěnová M, Borovička J, Kotrba P. Accumulation of Ag and Cu in *Amanita strobiliformis* and characterization of its Cu and Ag uptake transporter genes *AsCTR2* and *AsCTR3*. Biometals. 2016;**29**:249-264. DOI: 10.1007/s10534-016-9912-x

[78] Hložková K, Matěnová M, Žáčková P, Strnad H, Hršelová H, Hroudová M, Kotrba P. Characterization of three distinct metallothionein genes of the Ag-hyperaccumulating ectomycorrhizal fungus *Amanita strobiliformis*. Fungal Biology. 2016;**120**:358-369. DOI: 10.1016/j.funbio.2015.11.007.

[79] Mercogliano CP, DeRosier DJ. Gold nanocluster formation using metallothionein: mass spectrometry and electron microscopy. Journal of Molecular Biology. 2006;**355**:211-223. DOI: 10.1016/j.jmb.2005.10.026

[80] Mercogliano CP, DeRosier DJ. Concatenated metallothionein as a clonable gold label for electron microscopy. Journal of Structural Biology. 2007;**160**:70-82. DOI: 10.1016/j.jsb.2007.06.010

[81] Fernández de Castro I, Sanz-Sánchez L, Risco C. Metallothioneins for correlative light and electron microscopy. Methods in Cell Biology. 2014;**124**:55-70. DOI: 10.1016/B978-0-12-801075-4.00003-3

[82] Morphew MK, O'Toole ET, Page CL, Pagratis M, Meehl J, Giddings T, Gardner JM, Ackerson C, Jaspersen SL, Winey M, Hoenger A, McIntosh JR. Metallothionein as a clonable tag for protein localization by electron microscopy of cells. Journal of Microscopy. 2015;**260**:20-29. DOI: 10.1111/jmi.12262

[83] Presta A, Stillman MJ. Incorporation of copper into the yeast *Saccharomyces cerevisiae*. Identification of Cu(I)—metallothionein in intact yeast cells. Journal of Inorganic Biochemistry. 1997;**66**:231-240. DOI: 10.1016/S0162-0134(96)00216-4

[84] Nakayama K, Okabe M, Aoyagi K, Yamanoshita O, Okui T, Ohyama T, Kasai N. Visualization of yellowish-orange luminescence from cuprous metallothioneins in liver of Long-Evans Cinnamon rat. Biochimica et Biophysica Acta. 1996;**1289**:150-158

[85] McNulty M, Puljung M, Jefford G, Dubreuil RR. Evidence that a copper-metallothionein complex is responsible for fluorescence in acid-secreting cells of the *Drosophila* stomach. Cell and Tissue Research. 2001;**304**:383-389. DOI: 10.1007/s004410100371

[86] Ni TW, Staicu LC, Nemeth RS, Schwartz CL, Crawford D, Seligman JD, Hunter WJ, Pilon-Smits EA, Ackerson CJ. Progress toward clonable inorganic nanoparticles. Nanoscale. 2015;**7**:17320-17327. DOI: 10.1039/c5nr04097c

Chapter 3

Yeast as a Biocatalyst in Microbial Fuel Cell

Enas Taha Sayed and Mohammad Ali Abdelkareem

Additional information is available at the end of the chapter

http://dx.doi.org/10.5772/intechopen.70402

Abstract

Microbial fuel cells (MFCs) are fascinating bioelectrochemical devices that use the catalytic activity of living microorganisms to draw electric energy from organic matter present naturally in the environment or in the waste. Yeasts are eukaryotic microorganisms, classified as members of the fungus kingdom. Several yeast strains have been studied as biocatalysts in MFC with or without external mediator such as *Saccharomyces cerevisiae, Candida melibiosica, Hansenula anomala, Hansenula polymorpha, Arxula adeninvorans and Kluyveromyces marxianus*. In this chapter, we will focus on the use of yeast as a biocatalyst in the anode of microbial fuel cells (MFCs). How different yeast strains transfer electrons to the anode of the microbial fuel cells, advantages and challenges of the use of yeasts in MFCs, how to improve the performance and sustainability of the yeast-based MFCs through the modification of the anode electrode surface, and the application of the yeast-based MFCs in continuous wastewater treatment were discussed.

Keywords: yeast, microbial fuel cell, biocatalyst, electron transfer, mediator

1. Introduction

Microbial production of energy and/or chemicals from renewable carbohydrate feedstocks, and other organic-based wastes such as wastewater, is an attractive alternative to the current common fossil fuels. Microbial fuel cells (MFCs) are among the fast-growing microbial electrochemical systems (MESs) that offer a promising way for simultaneous wastewater treatment and electricity production [1–3]. Although MFCs showed promising features such as simultaneous wastewater treatment and electricity generation, low sludge production, wide range of substrates and operating at room temperature, the low power output and high cost especially that of the Pt cathode are the main challenges facing their commercialization [4–6].

In MFCs, the exo-electrogenic microorganisms act as biocatalysts in anaerobic oxidation of the organic materials that exist in different wastes, liberating electrons that can be collected by a conductive electrode, i.e., anode, generating an external power-producing circuit, and protons transferred through an electrolyte to a cathode surface. At the cathode, electrons react with protons and oxygen producing water [7–9]. The exo-electrogenic microorganisms that can be used in MFCs can be a prokaryote or eukaryote. Although prokaryotic microorganisms showed promising results in the MFCs and a lot of research has been carried out using them due to their ease in the electron transfer mechanism, yeast, as a eukaryote, attracted researchers' attention and was extensively studied as a biocatalyst in MFCs [4–6].

2. Microbial fuel cells: structure, components and mechanism

Microbial electrochemical systems (MESs) are innovative technology, recently implemented for numerous applications [10–15] such as (i) the simultaneous wastewater treatment and electricity production by MFCs, (ii) bio-hydrogen and/or other chemical production by microbial electrolysis cells (MECs), (iii) water desalination by microbial dialysis cells (MDCs) and (iv) electricity production in sediments or plant MFCs.

In case of MFCs, microorganisms oxidize organic matter, producing electrons that travel through a series of respiratory enzymes in the cell and make energy for the cell in the form of ATP. The electrons are then released to a terminal electron acceptor (TEA) that becomes reduced. Many TEAs such as oxygen, nitrate, sulfate and others readily diffuse into the cell where they accept electrons forming products that can diffuse out of the cell. However, it is now known that some microorganisms can transfer electrons exogenously (i.e., outside the cell) to a TEA such as metal oxides like iron oxide. This is the case of bacteria called exo-electrogens, which can be used to produce electricity in MFC [16].

Figure 1 shows a schematic diagram of an air-cathode MFC that consists of anode and cathode electrodes separated by a separator (if needed). The anode compartment composed of anode and carbon source (organic materials), with or without exogenous mediator. At the cathode, an electron acceptor (O_2 from air) reacts with protons that pass from the anode to the cathode through the electrolyte, and the electrons produce water.

2.1. Anode material

Anode material is considered as an important parameter that affects the performance of MFCs. The anode of the MFCs should have high electrical, mechanical and chemical stability, be biocompatible and have high surface area [20]. Carbon materials (conventional and nonconventional) are the best materials that are applied as anode in the MFCs showing high power output. The conventional carbon materials such as carbon paper, carbon cloth, carbon brush and carbon felt, and the nonconventional ones such as carbon nanotubes (CNTs), carbon nanofibers and graphene have been extensively applied in MFCs. Little work have been carried out using noncarbonaceous materials such as stainless steel, gold and titanium [17–19], which showed a lower performance compared to that obtained in case of using carbon.

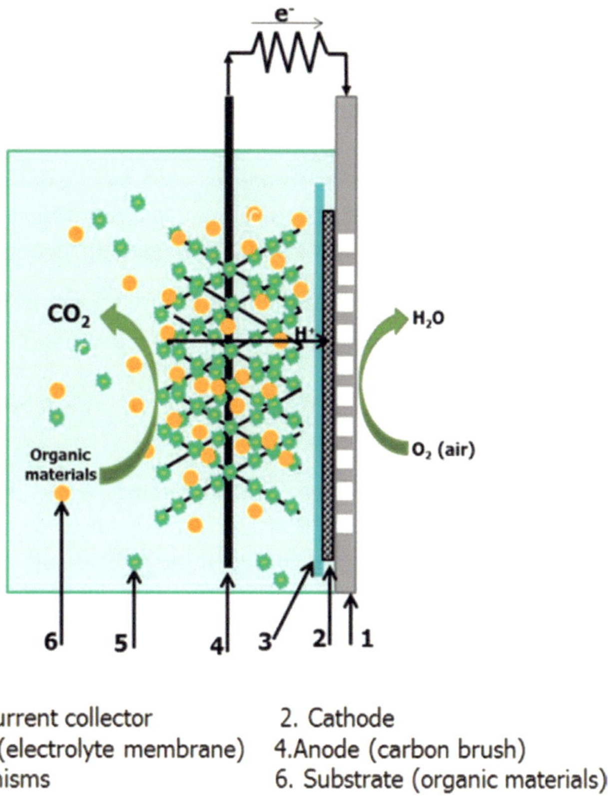

Figure 1. A schematic diagram showing the main components of an air-cathode MFC.

2.2. Cathode material

Cathode material has a significant impact on the overall cell voltage and it should have a high redox potential. Carbon materials such as carbon paper and carbon cloth modified with high active catalyst such as Pt catalyst are among the most common cathodes of the MFCs [20]. Although modifying the carbon cloth and/or carbon paper with Pt significantly decreased the oxygen reduction activation energy and increased the reaction rate, the high cost and scarcity of the Pt are the main challenges facing the application of such cathode. Recently, a wide range of non–Pt-based catalysts were investigated as cathodes in MFCs and showed promising results that gave them a potential to replace Pt catalyst in the near future such as carbon nitrogen alloys and metal carbides [18, 20–29].

2.3. Separator

As anode is working under anaerobic conditions, while cathode is working under aerobic conditions, the addition of separator with high ionic conductivity and low permeability could improve the MFC performance [30]. A large number of separators have been extensively studied in MFCs such as anion and cation exchange membranes, salt bridge, glass fibers, microfiltration membrane, porous fabrics, and coarse-pore filters [31–37]. It is worth mentioning that some MFCs showed better performance even without using the separator [3].

2.4. Microbes and electron transfer in microbial fuel cells

Microorganisms are generally divided into two main categories, prokaryotes and eukaryotes. Prokaryotes are simpler (no distinct nucleus) and smaller in size (around 1 μ in diameter) compared to eukaryotes that have larger size (5–10 μ or more) and are complex (possessing a distinct nucleus and subcellular organelles such as plastids and mitochondria) [4, 6]. All microorganisms that are capable of exo-cellular electron transfer (exo-electrogens) can be effectively used in MFCs without adding soluble exogenous mediators [4, 22, 30–38].

The possible electron transfer mechanisms in MFCs are shown in **Figure 2** and can be summarized in the following:

1. Direct electron transfer (DET) whether by direct cell attachment or through nanowires (pili)

   DET requires a direct contact between the anode surface and the outer membrane of the microorganism. Pili are nanowires that are formed out to connect the microorganism's membrane to the anode surface. The merits of the pili formation that multiple layers biofilm microorganisms can participate in the electron transfer while bulk ones do not participate in the electron transfer [4, 39–43].

2. Indirect electron transfer through external or internal mediators

   In this type, a redox active material (mediator) is responsible for the electron transfer between the microorganism and the anode surface. This redox can either be exerted naturally by the

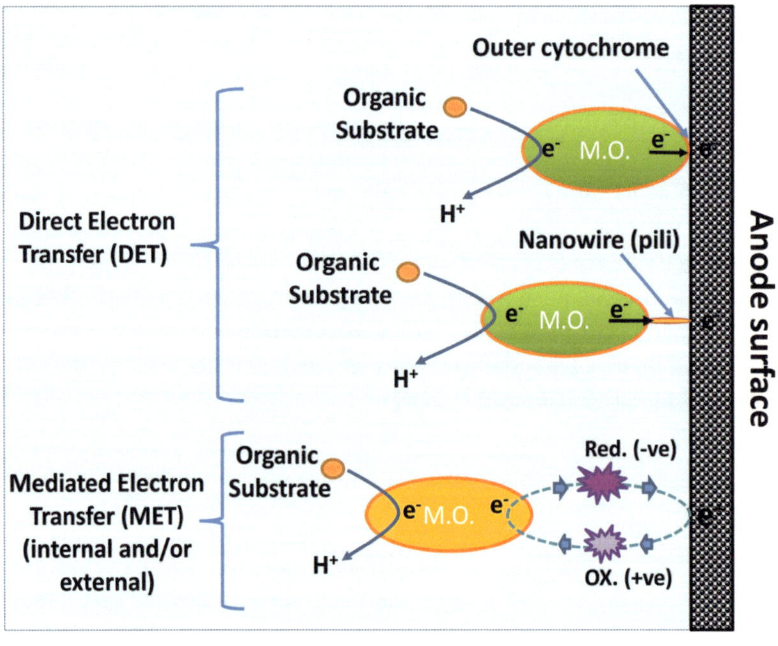

Figure 2. Schematic diagram showing different electron transfer mechanisms in MFCs.

microorganisms (internal) or can be added from outside (external). These mediators whether internal or external will be responsible for the electron transfer from the bulk microorganisms to the anode surface. The electron transfer in the mediated electron transfer is higher than that in the DET [4, 44–51].

Internal mediators have several advantages over the external ones such as they are cheap as they are exerted by the microorganism and have no toxic effect on the microorganism. **Figure 3** shows a schematic diagram of the disadvantages of external mediators and some types of the internal and external mediators.

Several external mediators have been investigated in MFCs such as methylene blue (MB), methyl red, methanyl yellow, methyl orange, bromocresol purple, bromocresol green (BcG), romothymol blue, bromophenol blue, Congo red, cresol red, eriochrome black T, murexide, neutral red (NR), yeast extract, etc.

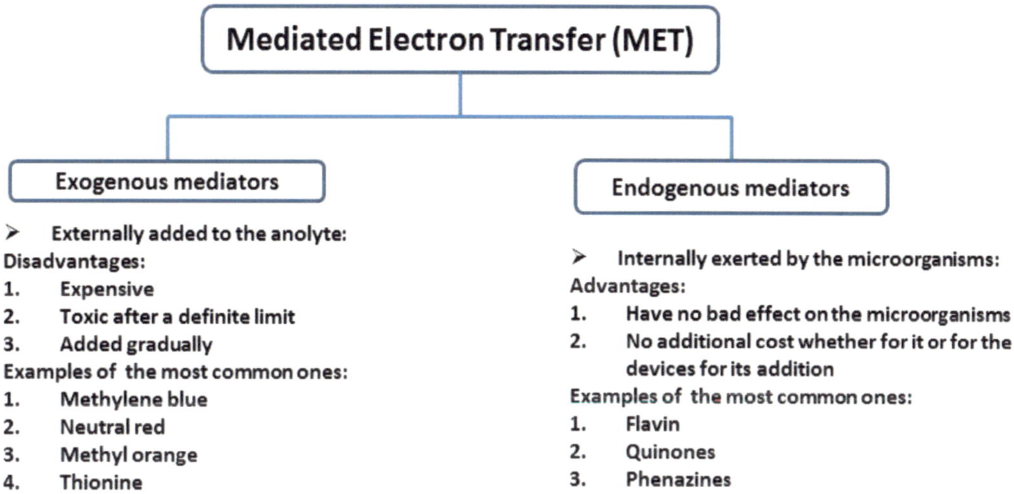

Figure 3. External and internal mediators in MFCs.

3. Yeast as a biocatalyst in MFCs

Yeast is a eukaryote with cell compartmentalization and has more complicated architecture compared to prokaryotes. Yeast is considered as an ideal biocatalyst for microbial fuel cell applications as most strains are nonpathogens, can metabolize wide range of substrates, are robust, and are easily handled. The bio-catalytic activity of the yeast would be related to the existence of different natural electron shuttles, mediators, such as azurin, ferredoxin and cytochromes, which could be used by redox enzymes for electron transfer from the yeast cells to the anode surface. This is in addition to the high extent of proteins in the yeast cell membrane, which is an important characteristic of electroactive species [4, 6]. Yeast cells also have a thick (100–200 nm) cell wall constructed of polysaccharides and proteins [43, 52]. Yeast cytochromes are located in the mitochondria, and transmembrane proteins (tPMETs) are located in the cell membrane, which are enclosed by the cell wall. Hence, to obtain an electrochemical response from the yeast cells, it has been assumed that a mediator must traverse the cell wall and interact with the membrane and/or internal redox sites such as NAD+/NADH [41, 42], or that the response originates from the soluble electroactive species exported from the cell [4, 45].

The electron transfer during the metabolism of the organic materials in the yeast cell is shown in **Figure 4**. Electrons liberate during the oxidation of the substrate into pyruvate in the glycolysis process, which takes place in the cytosol of the cell. These electrons received by the NAD+ forming NADH, which is recycled through its oxidation by the liberation of the electrons to the anode surface whether directly through the tPMETs or through the mediator to form NAD+ again — cycle of NADH to NAD^+. In mitochondria, oxidation of pyruvate into organic acids is associated with the liberation of the electrons that are received by the NAD+ forming NADH, which in turn are oxidized by releasing electrons to the mediator to form the NAD+ again. The reduced form of the mediator lost electrons to the anode surface to complete the cycle [38, 46].

Several yeast strains have been studied as biocatalysts in MFC with or without external mediator such as *Saccharomyces cerevisiae* (*S. cerevisiae*) [41–52], *Candida melibiosica* 2491 (*C. melibiosica*) [53–56], *Hansenula anomala* (*H. anomala*) [40], *Hansenula polymorpha* (*Hansenula polymorpha*) [57], *Arxula adeninivorans* (*A. adeninivorans*) [58] and *Kluyveromyces marxianus* (*K. marxianus*) [59].

3.1. *S. cerevisiae*

Baker's yeast (*S. cerevisiae*) is a single cell-based organism used in bread-making and beer production industry. *S. cerevisiae* is a simple eukaryotic cell, which serves as a model system

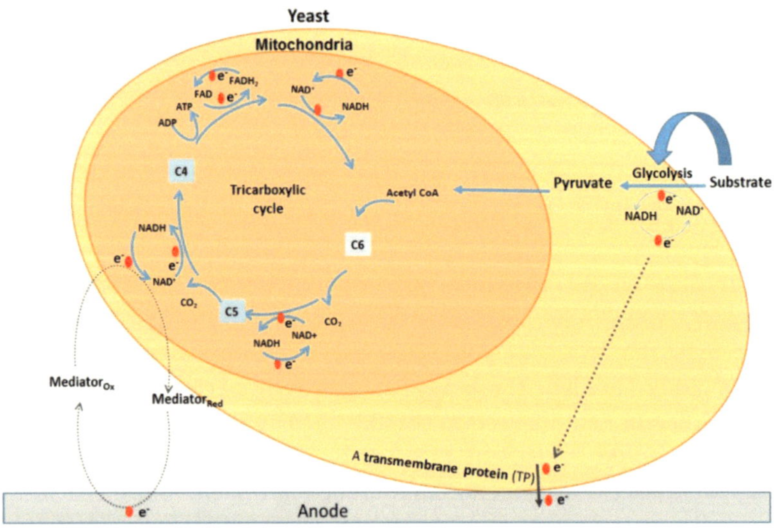

Figure 4. Schematic diagram shows the possible electrons' origin and transfer of yeast cells to MFC.

http://dx.doi.org/10.5772/intechopen.70402

[59, 60] for many eukaryotes, including human cells, for the study of fundamental cellular processes such as the cell cycle, DNA replication, recombination, cell division and metabolism [60, 61]. *S. cerevisiae* is considered to be a good biocatalyst in MFC due to its broad substrate spectrum, easy and fast mass cultivation, nonpathogenic, cheap and can be maintained for a long time in the dried state [9, 60–62]. Due to these features, *S. cerevisiae* was recently used in a large-scale MFC [63].

3.1.1. Mediator-less MFC

Mediator-less MFCs are those that operate without the addition of any external mediator. Sayed et al. [6] studied the mechanism by which *S. cerevisiae* transfers the electrons to the anode surface whether through the solution species or through the surface-confined species in a mediator-less MFC. *S. cerevisiae* was cultivated outside the MFC and then applied in an air-cathode mediator-less MFC using glucose as a substrate. Carbon paper was used as an anode and carbon paper with Pt/C as a cathode. When the MFC was operated with the yeast cells, the anode potential decreased from 0.4 to 0.1 V (vs. NHE pH 7) during 45 h. At the same time, the open circuit voltage (OCV) increased from 0.25 to 0.65 V. A maximum power output above 3 mW/m^2 was attained during the linear sweep voltammetry (LSV). At the end of the MFC operation, when the anolyte was replaced with a fresh one without yeast cells, i.e., just glucose, into the anode chamber, the cell attained the same maximum cell voltage within 1 h of cell operation. The same maximum power generation during the LSV was also attained. On the other hand, when another MFC using a fresh anode was operated with the filtered anolyte solution, i.e., no yeast cells, neither cell voltage nor anode potential changed, **Figures 5** and **6**. The ex-situ cyclic voltammetry of the filtered anolyte at the end of the experiments showed no redox peaks; i.e., no mediator existed in the anolyte. These measurements showed that the electron transfer was done through the surface-confined species, and there was no role of the solution species in it.

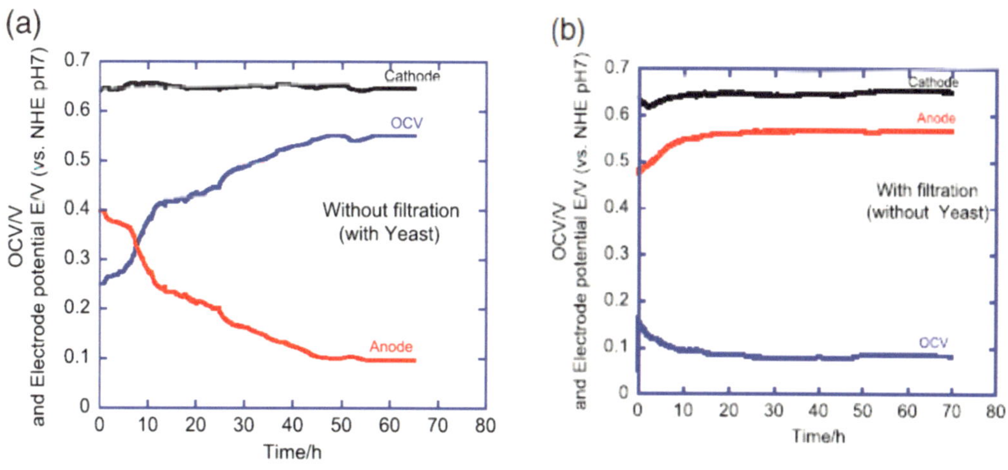

Figure 5. The OCV and the electrode potentials vs. time of the MFC using carbon paper (CP) as the anode material. (a) Without filtration and (b) with filtration [6].

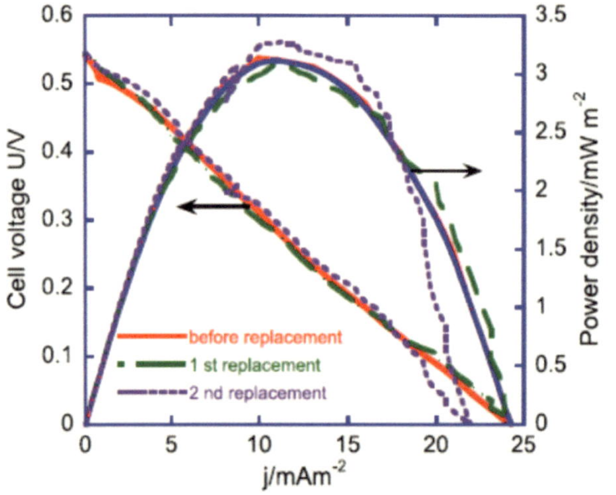

Figure 6. The i-V and i-p curves measured before and after the replacements of the anolyte solution [6].

The same conclusions for the direct electron transfer and no role of the mediator in the electron transfer of the *S. cerevisiae* were confirmed by Rawson et al. [41] who studied the direct electron transfer from the *S. cerevisiae* cells attached to the anode surface. The authors modified the anode surface with a mediator, osmium bipyridine complex, layer that hindered the mediator from penetrating the cell wall and reacting with the internal redox species. Results showed that the electron transfered from the yeast cells to the electrode surface through the yeast cell wall and no involvement of the endogenous mediator in this electron transfer.

In another study, the performance of air-cathode MFC using *S. cerevisiae* as an anodic biocatalyst under different redox conditions and organic loading was investigated [38]. The MFC was operated with synthetic wastewater at organic loading rate (OLR) of 0.91 kg COD/m^3-day and the performance of yeast-based MFC along with wastewater treatment was investigated at different feeding pH of 5.0, 6.0 and 7.0. Using cyclic voltammetry, which is an effective tool to identify the electron transfer mechanism in MFCs [64], the MFC performance was dependent on the OLR and the pH. Cyclic voltammetry confirmed the existence of the NADH/NAD+ and FADH/FAD+.

Although *S. cerevisiae* could be effectively used as a biocatalyst in mediator-less MFC, the power output was limited by the low electron transfer rate from the microorganism to the anode surface. The performance of the *S. cerevisiae*-based MFC could be improved by enhancing the rate of electron from the yeast cell to the anode surface by one or more of the following techniques:

i. Anode modification [42].

ii. Immobilization of the yeast cells on carbon nanotube [43].

iii. Yeast surface display of dehydrogenases [52].

iv. Addition of exogenous mediators [44–51].

3.1.1.1. *Enhancement of electron transfer in a mediator-less MFC*

The electrical conductivity of the anode plays an important role in the performance of the MFCs. The effect of the modification of carbon paper with thin layer of different transition metals, i.e., cobalt and gold, on the performance of air-cathode MFCs using *S. cerevisiae* as a biocatalyst was investigated [42]. Sputtering technique was used for preparing different thin layers of Co and Au with thicknesses of 5 and 30 nm on the surface of carbon electrodes. The 5-nm layer showed no significant effect on the cell performance, and this was related to the rare existence of the metals detected by the energy dispersive x-ray (EDX) measurements. On the other hand, 30 nm of Co significantly improved the performance where the power output increased from 12.8 to 20.2 mW/m^2 while the steady current discharge at 0.2 V increased from 8 to 27 mA/m^2, **Figure 7**. On the other hand, 30 nm of Au-modified electrode showed a negative effect on the cell performance. The positive effect of the Co on the performance was related to the enhancement of the electron transfer by the Co and the stimulation of the yeast growth on the modified electrode surface as confirmed by the SEM images. While Au suppressed the growth of the yeast cells as proved from the SEM images due to its poisoning effect, decreasing the performance (**Table 1**) [42].

The electron transfer of *S. cerevisiae* based MFC was enhanced by immobilizing *S. cerevisiae* on carbon nanotube (yeast/CNT) to be used as a catalyst in a membrane-less MFC [43]. The effect of the entrapping polymer (EP) and cross-linker (glutaraldehyde, GA) addition on the performance and stability of the MFC using *laccase* as cathodic catalyst was investigated. GA was selected as cross-liner due to its ability to promote cross-linking between yeast cells and poly(ethylenimine) (PEI), which used as the entrapping polymer due to its positive charge property. Bare CNT showed only C=C (sp^2) bonds indicating that CNT had not any functional group. In case of the immobilized yeast cells, C—N (C=N) bond peak appeared indicating that yeast cell and CNT were properly bonded. The immobilization of the yeast enhanced the power by 150% where it increased from 138 to 344 mW/m^2.

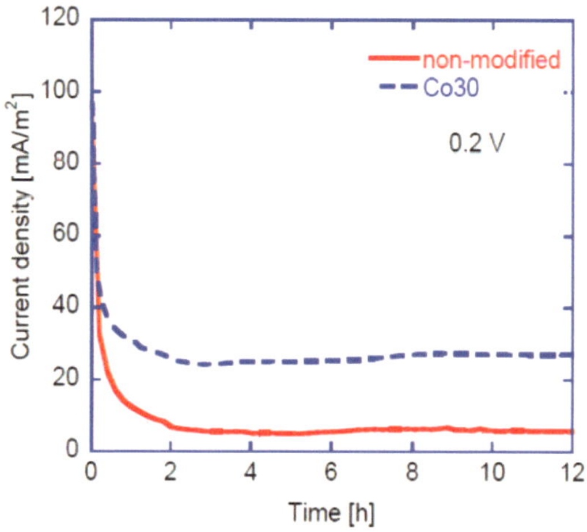

Figure 7. The i-t measurements at 0.2 V for a mediator-less yeast-based MFC using nonmodified (NME) carbon paper and Co, 30 nm, modified one [42].

Ref.	Max. power		Anode chamber (WV)	Separator	Cathode		Anode material	Carbon source	MFC type
	mW/m²	mW/m³			Electron acceptor	Electrode			
6	3.	17	84 mL (70 mL WV)	NRE 212	O_2 (air)	Pt/C over carbon paper	Carbon paper	Glucose	Air cathode
42	12.9 20.2 2		(70 mL WV)	Nafion 117	O_2 (air)	Pt/C over carbon paper	Carbon paper Co sputtered carbon paper Au-sputtered carbon paper	Glucose	Air cathode
38	25.51		350 mL (320 mL WV)	Nafion 117	O_2 (air)	Graphite plate	Graphite plate	Synthetic wastewater	Air cathode
52	2.7 2.8 33		8–10 mL (5 mL WV)	Nafion 117	O_2 (air)	A graphite plate	A graphite plate/ MWCNT	Lactose D-glucose lactose	Dual chamber
46		40	500 mL	Nafion 117	Potassium ferricyanide	Reticulated Vitreous carbon	Reticulated Vitreous carbon	Glucose	Dual chamber
47	28		850 mL (760 mL WV)	Nafion 117	-	Graphite plates	Graphite plates	Glucose	Dual chamber

Table 1. Summary of the studies done on the mediator-less *S. cerevisiae* yeast-based MFC.

The performance of *S. cerevisiae* based MFC was improved by displaying dehydrogenases, cellobiose dehydrogenase from *Corynascus thermophilus* (CtCDH) on the surface of *S. cerevisiae* using the yeast surface display system [52]. The surface displayed dehydrogenases were used in mediator-less two compartments MFCs. The MFCs were operated using unmodified *S. cerevisiae*, CtCDH-displaying *S. cerevisiae* and glucose oxidase (GOx) was used for comparison. Graphite plates modified with multi-walled carbon nanotubes (MWCNT) were used as electrodes in the anode and cathode compartments that were separated by Nafion 117. A maximum power output of CtCDH-displaying *S. cerevisiae* MFC was 33 mW/m^2 which was around 12 times higher than those obtained in case of GOx, and unmodified *S. cerevisiae*, 2.8 and, 2.7 mW/m^2, respectively.

3.1.2. Mediated yeast-based MFC

Several studies have been carried out to enhance the electron transfer through the addition of an external mediator. A candidate external mediator must satisfy several requirements such as being electrochemically active, fast release of electrons on the electrode surface, biocompatible to the microorganisms, soluble and chemically stable in the anolyte media, easily penetrate the cell membrane, and has a prober redox potential that is sufficiently positive to provide fast electron transfer from microorganisms to the anode while not too strong to avoid a big loss of potential [2, 14, 16]. Different mediators such as MB, NR, thionine, yeast extract, and others enhanced the electron transfer in *S. cerevisiae* yeast-based MFCs, and their power output are shown in **Table 2**.

Using copper electrodes and a sulfonated polyether ether ketone (SPEEK) as proton exchange membrane, Permana et al. [48] studied the performance of dual chamber *S. cerevisiae* yeast-based MFCs with and without MB using glucose as substrate. The MFC operated with MB showed higher cell voltage, higher power and energy outputs, and slightly lower glucose consumption without affecting the bioethanol production compared to the mediator-less MFC. Using rotating disc electrodes (RDEs), Ganguli and Dunn [45] were able to simultaneously determine the catalytic current under quiescent conditions along with the reduced mediator concentration that not adsorbed by the yeast. Based on the results from the anode kinetics study, a yeast powered microbial fuel cell successfully produced power density of ~1500 mW/m^2 once the reduced mediator concentration stabilized.

The effect of the anode modification on the performance of the mediated *S. cerevisiae* yeast-based MFC that used glucose as a substrate and MB as a mediator was investigated [50]. The anode carbon paper was sputtered with a thin layer of 30 nm of Co (Co30) or Au (Au30). The modification of the anode significantly improved the performance from 80 to 148 mW/m^2 and 120 mW/m^2 in case of Co30 and Au30, respectively, as shown in **Figure 8**. Although the cell resistance in case of Au is lower than that in Co, the performance of the latter was better and this was related to the poisons effect of the Au on the growth of the yeast cell on the anode surface; therefore, only the yeast in the anolyte took part in the performance, while in case of the Co, the yeast cells in the anolyte and those formed as biofilm on the anode surface took part in the electron transfer. The cell resistance decreased from 25 $\mu\Omega$ cm^2 in the case of nonmodified (NME) anode to 4 and 3 $\mu\Omega$ cm^2 in case of Co30 and Au30, respectively. The better performance

Ref.	Max. power		Mediator	Anode chamber (WV)	Separator	Cathode		Anode material	Carbon source	MFC type
	mW/m²	mW/m³				Electron acceptor	Electrode			
59	22	850×10^3	2-hydroxy-1,4-naphthoquinone	WV, 7.5 cm³	Gore-Tex, 30 μm	$K_3[Fe(CN)_6]$	Carbon rods	Carbon rods and carbon fiber bundles	Glucose	Dual-chamber
	80							Carbon paper		
50	148		MB	(70 mL WV)	Nafion 117	O_2 (air)	Pt/C over carbon paper	Co-sputtered carbon paper	Glucose	Air cathode
	120							Au-sputtered carbon paper		
45	150		MB	10 mL	Nafion	Potassium ferricyanide	Carbon felt	Carbon felt	Glucose	Dual chamber
46		146.71 ± 7.7	MB	500 mL	Nafion 117	Potassium ferricyanide	Reticulated vitreous carbon	Reticulated vitreous carbon	Glucose	Dual chamber
	39								D-xylose	
	31								D-glucose	
52	32		MB (0.1 M)	25 mL		O_2 (air)	Pt/C over carbon cloth	Graphite plate	L-arabinose	Air cathode
	22								D-cellobiose	
	14								D-galactose	
	400		MB							
44	80		NR	32 mL	Nafion 115		Reticulated vitreous carbon	Reticulated vitreous carbon	Dextrose	Dual chamber
	500		MB &NR							

Ref.	Max. power		Mediator	Anode chamber (WV)	Separator	Cathode		Anode material	Carbon source	MFC type
	mW/m^2	mW/m^3				Electron acceptor	Electrode			
45	1500		MB	10 mL	Nafion	Potassium ferricyanide	Carbon felt,	Carbon felt	Glucose	Dual chamber MFC
46		145	MB	500 mL	Nafion 117	Potassium ferricyanide	Reticulated vitreous carbon,	Reticulated vitreous carbon	Glucose	Dual chamber MFC
47				850 mL (760 mL WV)	Nafion 117	-	Graphite plates	Graphite plates		Dual chamber
51	36 70	36	YE	70 mL WV	Nafion 117		Pt/C over carbon paper	Carbon paper Au-plated carbon paper	Glucose	Air cathode

Table 2. Summary of the studies done on the mediated *S. cerevisiae* yeast-based MFC.

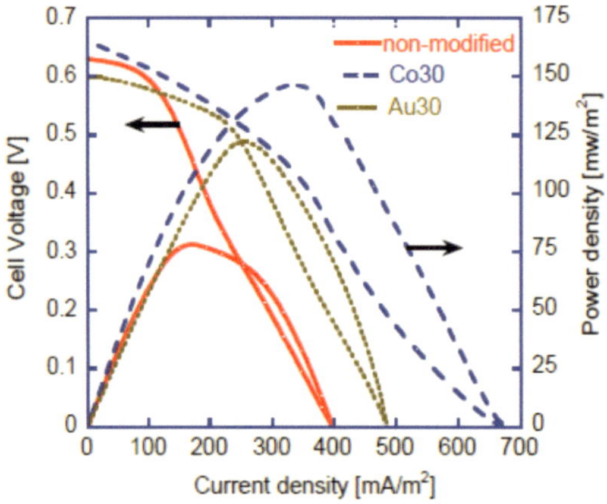

Figure 8. The i-V and i-p curves of the yeast-based MFC with 0.1 mM MB using nonmodified carbon paper, and Co30 and Au30 as anodes [50].

in both cases was related to the metal-modified surface that significantly enhanced the electron transfer via the exogenous mediator. It was also considered that the highly conductive surface of the Co or Au on the anode surface increased the efficiency of the electron transfer by contacting a part of the mediator with an electric charge on the anode.

MB was also used in air-cathode MFC that used modified *S. cerevisiae* using yeast surface display system [52]. Pyranose dehydrogenase from *Agaricus meleagris* (AmPDH) was displayed on the surface of *S. cerevisiae*. The MFCs were operated using unmodified *S. cerevisiae* or AmPDH-displaying *S. cerevisiae* with various fuels, D-xylose, D-glucose, L-arabinose, D-cellobiose and D-galactose using 0.1 mM MB. AmPDH displaying *S. cerevisiae* generated high power outputs using the different substrates, 3.1, 3.9, 3.2, 2.2, and 1.4 μW/cm^2 in case of using D-glucose, D-xylose, L-arabinose, D-cellobiose and D-galactose, respectively, compared with a maximum power output of 0.8 μW/cm^2 in case of the unmodified *S. cerevisiae* using D-xylose as a fuel [52].

Compared to MB, NR showed promising results in a two-compartment *S. cerevisiae* yeast-based MFC for degradation of whey. With a fixed concentration of the two mediators of 100 μmol/l, the maximum power and current densities increased from 1.43 μW and 11.5 μA to 50 μW and 470 μA in case of the NR compared to 11.3 μW and 120 μA in case of MB. These results showed that NR served as a suitable mediator and enhanced the electrical energy by 5 folds compared to that of MB [49]. When NR (0.5 mM) was added to the MB (0.5 mM), the *S. cerevisiae* yeast-based MFC showed a maximum power output of 500 mW/m^2 compared to 400 mW/m^2 in case of 1 mM of MB [44]. This increase in the performance was related to the role of the MB in the enhancement of the anaerobic respiration, while NR involved with fermentation only. This study [44] showed that the addition of the MB was effective than NR, which is in contradiction to that reported by Najafpour et al. [49]. This might be related to the difference in the operation conditions, and/or any other reasons that is not clear for the authors right now.

Thionine is another mediator that worked effectively in *S. cerevisiae* yeast-based MFC [47]. Thionine addition significantly increased the performance from 3 to 28 mW/m^2. An optimum concentration of thionine was 500 mM, giving a maximum voltage of 420 mV and a maximum current of 700 mA/m^2. Cyclic voltammetry measurements showed a redox peak of −0.1 V vs. Ag/AgCl.

Yeast extract, which is one of the main components of the biological cultivating media, was effectively used as a mediator in *S. cerevisiae* yeast-based MFC [51]. Using two different anodes, plain carbon paper and gold-plated carbon paper, the current density increased from 94 and to 190 and 300 mA/m^2, respectively, by yeast extract addition as shown in **Figures 9a** and **10a**. While the power density increased from 12.9 and 2 to 32.6 and 70 mW/m^2 with the yeast extract addition for the plain and the gold-plated electrodes, respectively, as shown in **Figures 9b** and **10b**. The role of the yeast extract as an electron transfer mediator was confirmed using the gold-plated carbon paper where no cells were detected on its surface (as confirmed from the scanning electron microscopic [SEM] images); therefore, the role of the surface-confined species in the cell performance was denied.

3.2. *C. melibiosica*

C. melibiosica is a yeast strain that possess high phytase activity, which existed in plant wastes. This yeast strain was used in numerous studies as a biocatalyst in MFC with and without mediator as can be seen in **Table 3**. The catalytic activity of *C. melibiosica* was studied in a dual chamber MFC with and without the addition of MB using different carbon sources, i.e., fructose, glucose and sucrose [53]. Results showed that *C. melibiosica* could be used as a biocatalyst in a mediator-less MFC giving a maximum power output of 60 mW/m^3 in case of fructose. This power increased three times either by the addition of yeast extract and peptone or by the MB addition [53].

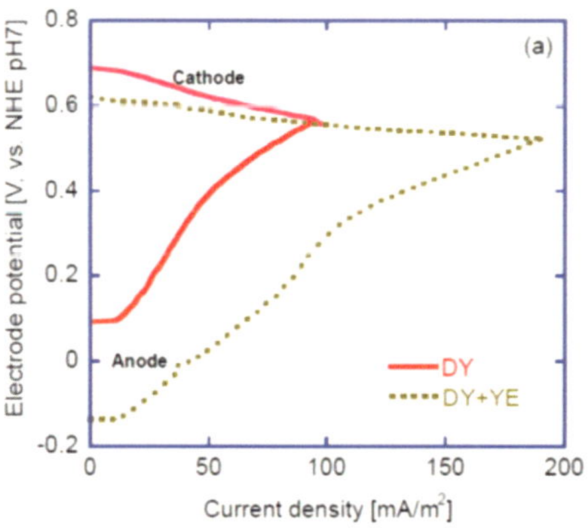

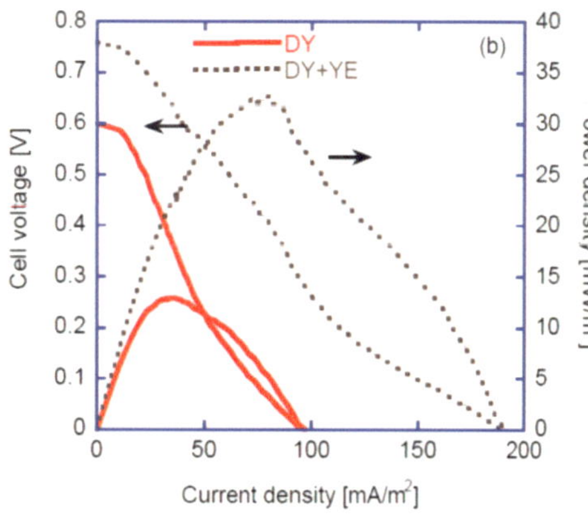

Figure 9. The effect of the yeast extract (YE) addition to *S. cerevisiae* (DY)-based MFC using nonmodified carbon paper as anode on (a) the electrode potentials, and (b) the current-voltage and current-power curves [51].

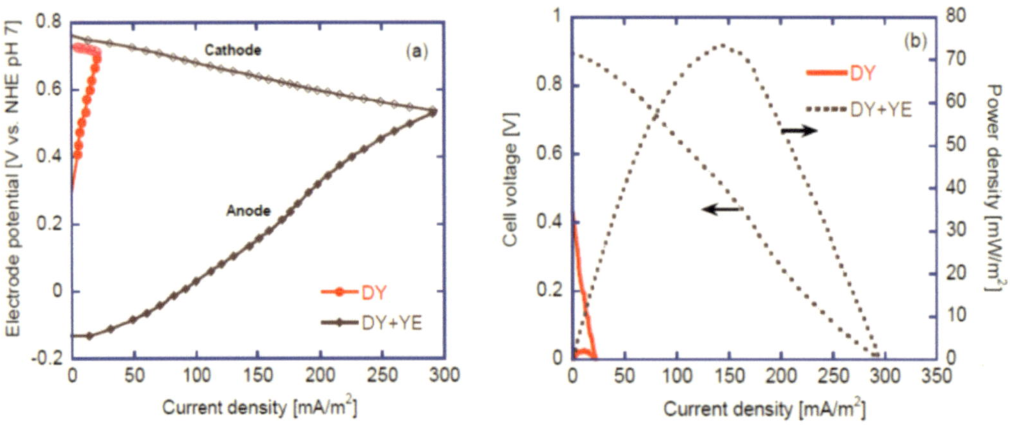

Figure 10. The effect of the yeast extract (YE) addition on *S. cerevisiae* (DY)-based MFC in case of using gold-sputtered carbon paper as anode on (a) the electrode potentials, and (b) the current-voltage and current-power curves [51].

The effect of the mediator type, i.e., bromocresol green (BcG), bromocresol purple, romothymol blue, bromophenol blue, Congo red, cresol red, eosin, eriochrome black T, methyl red, methanyl yellow, MB, methyl orange, murexide and NR on the performance of *C. melibiosica*-based MFC was investigated [54]. Results showed that among the investigated mediators, MB, methyl orange, methyl red and NR increased the performance compared to the mediator-less MFC. MB showed the best among all of them where the performance increased from 20 to 640 mW/m^2 with MB concentration of 0.8 mM. This was related to its ability not only to increase the electron transfer rate but also forcing the living cells to switch on various catabolic pathways and divert electrons from different energetic levels, thus increasing the energy production. This had been confirmed by measuring the ethanol production. where the MFC that operated using MO and MR produced trace amounts of ethanol, while in case of MB, ethanol was not detected. These indicated that the aerobic respiration processes were predominant in these cases. On the other hand, ethanol was produced in large quantities when NR and BcG were used, demonstrating that these mediators stopped the respiratory processes and displaced them with alcoholic fermentation.

The performance of *C. melibiosica*-based MFC was investigated using modified and nonmodified (NME) carbon felt [55]. The carbon felt was modified by Ni using two different techniques, i.e., galvanostatic pulse (GME) and potentiostatic pulse (PME). Carbon felt was used as the cathode, Nafion 117 as the separator, fructose, yeast extract and peptone (YP_{fru}) as the anolyte, and potassium ferricyanide as the catholyte. The power output of the MFC significantly increased using the modified electrodes where it increased from 36 mW/m^2 in case of the NME to 390 and 720 mW/m^2 in case of PME and GME, respectively. These values were even higher than that obtained in case of using the NME with addition of the external mediator, MB. The authors related the improvement in the cell performance to the existence of Ni ions, which acted as an electron acceptor and/or due to adaptive mechanism which enhanced electron transfer through the yeast membrane. In another study, the authors prepared carbon felt modified with NiFe and NiFeP using the same preparation method [56]. They found

http://dx.doi.org/10.5772/intechopen.70402

Ref.	Max. power		Electron transfer mechanism		Anode chamber (WV)	Separator	Cathode		Anode material	Carbon source	MFC type
	mW/m²	mW/m³	Mediator less	Mediator			Electron acceptor	Electrode			
		60	Mediator less							Fructose,	
53		180	Mediator less		100 mL	Salt bridge	Potassium ferricyanide	Graphite rods,	Graphite rods	YP_{fru}	Dual chamber
		185		MB						Fructose	
54	640			MB	13 mL	Nafion 117	Potassium ferricyanide	Carbon felt	Carbon felt	YP_{fru}	Two chamber
	36								Carbon felt NME		
55			Mediator less		13 mL	Nafion 117	Potassium ferricyanide	Carbon felt		Fructose	Dual chamber
	720								Ni-nanomodified carbon felts galvanostatic pulse deposition (GME)		
	390								Ni-nanomodified carbon felts potentiostatic pulse technique (PME)		
56	83		Mediator less		13 mL	Nafion 117	Potassium ferricyanide	Carbon felt	NiFe(g.)	Fructose	Dual chamber
	93								NiFe(p.)		
	155								NiFeP(g.)		

Table 3. Summary of the studies done on the *C. melibiosica* yeast-based MFC.

that among the different tested electrodes, NiFeP-modified electrodes showed the best performance of 260 ± 8 and 155 ± 6 mW/m^2 prepared potentiostatically and galvanostatically, respectively. The authors related the improvement in the performance to same reasons that were described above [55].

3.3. Other yeast strains

3.3.1. *H. anomala*

The catalytic activity of *H. anomala* in a mediator-less MFC using glucose as the substrate was investigated [40]. The *H. anomala* cells were immobilized on the surface of the anode by physical adsorption and covalent linkage. The results showed that *H. anomala* could transfer the electrons through the redox proteins, i.e., *ferricyanide reductase* and *lactate dehydrogenase* exist in their outer membrane. Moreover, the MFC was operated using different anodes, i.e., graphite, graphite felt and polyaniline–Pt-composite-coated graphite. A maximum power output of 2.34, 2.9 and 0.69 W/m^3 was obtained in case of graphite felt, graphite modified with PANI and Pt, and graphite, respectively. The high performance was related to the high surface of the graphite felt and the presence of the catalytic active Pt in case of the graphite modified with PANI and Pt, respectively.

3.3.2. *H. polymorpha*

The electron transfer pathways between the cytosolic redox enzymes of *H. polymorpha*, overexpressing flavocytochrome b2 (FC b2), and the electrode surface was studied [57]. Both wild and genetic *H. polymorpha* yeast cells were entrapped in osmium-complex-modified redox polymers (OsRP), which are essential for the electron transfer communication, on the surface of graphite electrodes. With the addition of L-lactate, current generation was noticeable when genetic modified one was used and it was in direct contact with the redox polymer, i.e., OsRP. The results suggested that the overexpression of FC b2 and the related amplification of the FC b2/ L-lactate reaction cycle were essential to provide enough charge to the electron-exchange network in order to facilitate sufficient electrochemical coupling between the cells, via the redox polymer, and the electrode. Also they suggested that the intimate contact between the cell walls and the redox polymer is a prerequisite for electrically wiring the cytosolic FC b2/ L-lactate redox activity.

3.3.3. *A. adeninvorans*

The biocatalytic activity of the nonconventional yeast *A. adeninvorans* in a mediator-less dual chamber MFC was investigated [58]. Results showed that *A. adeninvorans* was effectively used as a biocatalyst ion in the MFC, generating a power of more than 0.025 W/m^2. The electron transfer was confirmed to be through the secretion of an endogenous mediator in the solution. This was confirmed using cyclic voltammetry of the supernatant from the *A. adeninvorans*. An irreversible oxidation peak at +0.45 V appeared. An *A. adeninvorans* yeast-based MFC showed a better performance than that obtained in case of *S. cerevisiae* yeast-based MFC, and this was related to the exertion of endogenous mediator in case of *A. adeninvorans*.

3.3.4. *K. marxianus*

Kaneshiro et al. [59] have investigated the catalytic activity of six different yeast strains in a dual chamber MFC with glucose as the substrate including *K. marxianus, S. cerevisiae, Pichia pastoris, H. polymorpha, Kluyveromyces lactis, Schizosaccharomyces pombe, Candida glabrata* and yeast strains isolated from soil [59]. Among the different tested yeast strains, *K. marxianus* showed the highest cell performance followed by *S. cerevisiae* and *P. pastoris*. Although *K. marxianus* showed the lowest glucose consumption, it showed the lowest ethanol production indicating highest efficiency. Furthermore, *K. marxianus* showed catalytic activity for the metabolism of fructose and xylose; therefore, the authors suggested that *K. marxianus* could be effectively used for of woody biomass. *K. marxianus* is one of the robust yeast strains that could be used at high temperature; therefore, the authors investigated its catalytic activity under different temperatures, 37, 45 and 50°C. The results showed that *K. marxianus* had the highest activity at 45°C. This could be used for the treatment of high-temperature effluents that are produced in some industries.

4. Large-scale yeast-based MFC

A novel yeast-based MFC stack that composed of 4 units of total capacity of 1840 mL was designed and operated using glucose as the carbon source, graphite plates as the electrodes and Nafion 117 as the separator [63]. The stack was operated under continuous mode with a hydraulic retention time of 6.7 h. Single cell and cells connected in parallel and/or series connections were investigated to achieve the best operating conditions. A maximum current of 6447 mA/m^2 and maximum power of 2003 mW/m^2 were obtained. A Columbic efficiency of 22% was obtained in the parallel connection. **Figure 11** showed that the stack could be operated for more than 3 days with stable voltage and power output. The results obtained in this study proved the potential of yeast for scaling up. **Table 4** showed summary of the materials and operating conditions used in the stack.

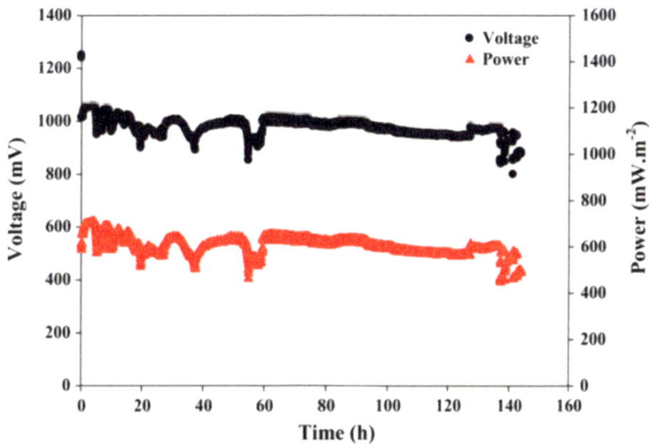

Figure 11. Close circuit voltage and produced power from staked MFC at parallel mode with 1 KΩ resistances in external circuit for 148 h [63].

MFC material	Plexiglas
MFC type	MFCs stack composed of 4 anodes and 3 cathodes compartments
Anode	Graphite plates, size of 40 × 60 × 1.2 mm
Cathode	Graphite plates, size of 40 × 60 × 1.2 mm
Membrane	Nafion 117.32 cm^2
Catholyte	Potassium permanganate (400 μmol/L)
Anode media	Yeast (*S. cerevisiae* PTCC 5269). NR (200 μmol/L)
Fuel	Glucose, 30 g/L
Anode chamber (volume)	460 mL
Working volume	350 mL
Current collector	Copper wire
Mode	Continuous up flow mode
HRT	6.7 h

Table 4. A summary of the stack materials and operating conditions.

5. Conclusions and recommendations

Yeast is successfully used as a biocatalyst in MFC, which exhibits different electron transfer mechanisms according to its strains. In *S. cerevisiae* and *H. anomala*, the electron transfer takes place through the surface-confined species; in *C. melibiosica, H. polymorpha and A. adeninivorans*, the transfer of electrons from yeast cells to the anode is both by the secretion of redox molecules and by the direct electron transfer. The modification of the anode and the addition of external mediator significantly enhanced the cell performance. *K. marxianus* is one of the most promising yeast strains as it could effectively metabolize the complex organic materials with high power output even under high operating temperature conditions; therefore, it could be a best choice for wastes with fluctuated temperature. Further studies on this type and other types are required. Moreover, the surface modification of the carbon material with graphene could improve the performance.

Author details

Enas Taha Sayed[1*] and Mohammad Ali Abdelkareem[1,2]

*Address all correspondence to: e.kasem@mu.edu.eg

1 Chemical Engineering Department, Faculty of Engineering, Minia University, El-Minya, Egypt

2 Department of Sustainable and Renewable Energy Engineering, University of Sharjah, Sharjah, United Arab Emirates

References

[1] Rabaey K, Lissens G, Siciliano S, Verstraete W. A microbial fuel cell capable of converting glucose to electricity at high rate and efficiency. Biotechnology Letters. 2003;**25**:531-1535

[2] Tender L, Reimers C, Stecher H, Holmes D, Bond D, Lowy D, Pilobello K, Fertig S, Lovley D. Harnessing microbially generated power on the seafloor. Nature Biotechnology. 2002;**20**:821-825

[3] Liu H, Logan BE. Electricity generation using an air-cathode single chamber microbial fuel cell in the presence and absence of a proton exchange membrane. Environmental Science & Technology. 2004;**38**:4040-4046

[4] Schaetzle O, Barrie're F, Baronian K. Bacteria and yeasts as catalysts in microbial fuel cells: Electron transfer from micro-organisms to electrodes for green electricity. Energy & Environmental Science. 2008;**1**:607-620

[5] Gunawardena A, Fernando S, To F. Performance of a yeast-mediated biological fuel cell. International Journal of Molecular Sciences. 2008;**9**:1893-1907

[6] Sayed E, Tsujiguchi T, Nakagawa N. Catalytic activity of baker's yeast in a mediator-less microbial fuel cell. Bioelectrochemistry. 2012;**86**:97-101

[7] Min B, Logan BE. Continuous electricity generation from domestic wastewater and organic substrates in a flat plate microbial fuel cell. Environmental Science & Technology. 2004;**38**:5809-5814

[8] He Z, Minteer S, Angenent L. Electricity generation from artificial wastewater using an upflow microbial fuel cell. Environmental Science & Technology. 2005;**39**:5262-5267

[9] He L, Du P, Chen Y, Lu H, Cheng X, Chang B, Wang Z. Advances in microbial fuel cells for wastewater treatment. Renewable and Sustainable Energy Reviews. 2017;**71**:388-403

[10] Harnisch F, Schrooder U. From MFC to MXC: Chemical and biological cathodes and their potential for microbial bioelectrochemical systems. Chemical Society Reviews. 2010;**39**:4433-4448

[11] Hays S, Zhang F, Logan BE. Performance of two different types of anodes in membrane electrode assembly microbial fuel cells for power generation from domestic wastewater. Journal of Power Sources. 2011;**196**:8293-8300

[12] Jeremiasse AW, HVM H, CJN B. Microbial electrolysis cell with a microbial biocathode. Bioelectrochemistry. 2010;**78**:39-43

[13] Cao X, Huang X, Liang P, Xiao K, Zhou Y, Zhang X, Logan BE. A new method for water desalination using microbial desalination cells. Environmental Science & Technology. 2009;**43**:7148-7152

[14] Chiranjeevi P, Mohanakrishna G, Mohan SV. Rhizosphere mediated electrogenesis with the function of anode placement for harnessing bioenergy through CO_2 sequestration. Bioresource Technology. 2012;**124**:364-370

[15] ElMekawy A, Hegab HM, Vanbroekhoven K, Pant D. Techno-productive potential of photosynthetic microbial fuel cells through different configurations. Renewable and Sustainable Energy Reviews. 2014;**39**:617-627

[16] Logan BE, Microbial Fuel Cells. 1st ed. Wiley; 2007

[17] Richter H, McCarthy K, Nevin K, Johnson J, Rotello V, Lovley D. Electricity generation by Geobacter sulfurreducens attached to gold electrodes. Langmuir. 2008;**24**:4376-4379

[18] Dumas C, Mollica A, Feron D, Basseguy R, Etcheverry L, Bergel A. Marine microbial fuel cell: Use of stainless steel electrodes as anode and cathode materials. Electrochimica Acta. 2007;**53**:468-473

[19] Heijne A, Hamelers H, Saakes M, Buisman C. Performance of non-porous graphite and titanium-based anodes in microbial fuel cells. Electrochimica Acta. 2008;**53**:5697-5703

[20] Watanabe K. Recent developments in microbial fuel cell technologies for sustainable bioenergy. Journal of Bioscience and Bioengineering. 2008;**6**:528-536

[21] Wang B. Recent development of non-platinum catalysts for oxygen reduction reaction. Journal of Power Sources. 2005;**152**:1-15

[22] Logan BE, Hamelers B, Rozendal R, Schro¨der U, Keller J, Freguia S, et al. Microbial fuel cells: Methodology and technology. Environmental Science & Technology. 2006; **40**(51):81-92

[23] Chen Z, Higgins D. Nitrogen doped carbon nanotubes and their impact on the oxygen reduction reaction in fuel cells. Carbon. 2010;**48**:57-65

[24] Roche I, Katuri K, Scott K. A microbial fuel cell using manganese oxide oxygen reduction catalysts. Journal of Applied Electrochemistry. 2010;**40**:13-21

[25] Zhang L, Liu C, Zhuang L, Li W, Zhou S, Zhang J. Manganese dioxide as an alternative cathodic catalyst to platinum in microbial fuel cells. Biosensors & Bioelectronics. 2009;**24**:2825-2829

[26] Lu M, Kharkwal S, Ng H, Li S. Carbon nanotube supported MnO_2 catalysts for oxygen reduction reaction and their applications in microbial fuel cells. Biosensors & Bioelectronics. 2011;**26**:4728-4732

[27] Zhang Y, Hu Y, Li S, Sun J, Hou B. Manganese dioxide-coated carbon nanotubes as an improved cathodic catalyst for oxygen reduction in a microbial fuel cell. Journal of Power Sources. 2011;**196**:9284-9289

[28] Liu XW, Sun XF, Huang YX, Sheng GP, Zhou K, Zeng RJ, et al. Nano-structured manganese oxide as a cathodic catalyst for enhanced oxygen reduction in a microbial fuel cell fed with a synthetic wastewater. Water Research. 2010;**44**:5298-5305

[29] Wen Q, Wang S, Yan J, Cong L, Pan Z, Ren Y, et al. MnO_2-Graphene hybrid as an alternative cathodic catalyst to platinum in microbial fuel cells. Journal of Power Sources. 2012;**216**:187-191

[30] Kim I, Chae K, Choi M, Verstraete W. Microbial fuel cells: Recent advances, bacterial communities and application beyond electricity generation. Environmental Engineering Research. 2008;**13**:51-65

[31] Harnisch F, Warmbier R, Schneider R, Schroder U. Modeling the ion transfer and polarization of ion exchange membranes in bioelectrochemical systems. Bioelectrochemistry. 2009;**75**:136-141

[32] Oh S, Logan BE. Proton exchange membrane and electrode surface areas as factors that affect power generation in microbial fuel cells. Applied Microbiology and Biotechnology. 2006;**70**:162-169

[33] Kim J, Cheng S, Eun OS, Logan BE. Power generation using different cation, anion, and ultrafiltration membranes in microbial fuel cells. Environmental Science & Technology. 2007;**41**:1004-1009

[34] Sun J, Hu Y, Bi Z, Cao Y. Improved performance of air-cathode single-chamber microbial fuel cell for wastewater treatment using microfiltration membranes and multiple sludge inoculation. Journal of Power Sources. 2009;**187**:471-479

[35] Mohan S, Raghavulu S, Sarma P. Biochemical evaluation of bioelectricity production process from anaerobic wastewater treatment in a single chambered microbial fuel cell (MFC) employing glass wool membrane. Biosensors & Bioelectronics. 2006;**23**:1326-1332

[36] Zhang X, Cheng S, Wang X, Huang X, Logan BE. Separator characteristics for increasing performance of microbial fuel cells. Environmental Science & Technology. 2009; **43**:8456-8461

[37] Min B, Cheng S, Logan BE. Electricity generation using membrane and salt bridge microbial fuel cells. Water Research. 2005;**39**:1675-1686

[38] Raghavulu V, Goud RK, Sarma PN, Mohan SV. *S. cerevisiae* as anodic biocatalyst for power generation in biofuel cell: Influence of redox condition and substrate load. Bioresource Technology. 2001;**102**:2751-2757

[39] Schröder U. Anodic electron transfer mechanisms in microbial fuel cells and their energy efficiency. Physical Chemistry Chemical Physics. 2007;**9**:2619-2629

[40] Prasad D, Arun S, Murugesan M, Padmanaban S, Satyanarayanan RS, Berchmans S, Yegnaraman V. Direct electron transfer with yeast cells and construction of a mediatorless microbial fuel cell. Biosensors and Bioelectronics. 2007;**22**:2604-2610

[41] Rawson F, Gross A, Garrett D, Downard A, Baronian K. Mediated electrochemical detection of electron transfer from the outer surface of the cell wall of *S. cerevisiae*. Electrochemistry Communications. 2012;**15**:85-87

[42] Kasem E, Tsujiguchi T, Nakagawa N. Effect of metal modification to carbon paper anodes on the performance of yeast-based microbial fuel cells part I: In the case without exogenous mediator. Key Engineering Materials. 2013;**534**:76-81

[43] Christwardana M, Kwon Y. Yeast and carbon nanotube based biocatalyst developed by synergetic effects of covalent bonding and hydrophobic interaction for performance enhancement of membraneless microbial fuel cell. Bioresource Technology. 2017;**225**:175-182

[44] Wilkinson S, Klar J, Applegarth S. Optimizing biofuel cell performance using a targeted mixed mediator combination. Electroanalysis. 2006;**18**:2001-2007

[45] Ganguli R, Dunn BS. Kinetics of anode reactions for a yeast-catalysed microbial fuel cell. Fuel Cell. 2009;**9**:44-52

[46] Gunawardena A, Fernando S, To F. Performance of a yeast-mediated biological fuel cell. International Journal of Molecular Sciences. 2008;**9**:1893-1907

[47] Rahimnejad M, Najafpour G, Ghoreyshi A, Talebnia F, Premier G, Bakeri G, Kim J, Oh S. Thionine increases electricity generation from microbial fuel cell using *S. cerevisiae* and exoelectrogenic mixed culture. The Journal of Microbiology. 2012;**50**:575-580

[48] Permana D, Rosdianti D, Ishmayana S, Rachman S, Putra H, Rahayuningwulan D, Hariyadi H. Preliminary investigation of electricity production using dual chamber microbial fuel cell (DCMFC) with *S. cerevisiae* as biocatalyst and methylene blue as an electron mediator. Procedia Chemistry. 2015;**17**:36-43

[49] Najafpour G, Rahimnejad M, Mokhtarian N, Daud W, Ghoreyshi A. Bioconversion of whey to electrical energy in a biofuel cell using *S. cerevisiae*. World Applied Sciences Journal. 2010;**8**:01-05

[50] Kasem E, Tsujiguchi T, Nakagawa N. Effect of metal modification to carbon paper anodes on the performance of yeast-based microbial fuel cells part II: In the case with exogenous mediator, methylene blue. Key Engineering Materials. 2013;**534**:82-87

[51] Sayed ET, Barakat NAM, Abdelkareem MA, Fouad H, Nakagawa N. Yeast extract as an effective and safe mediator for the Baker's-yeast-based microbial fuel cell. Industrial and Engineering Chemistry Research. 2015;**54**:3116-3122

[52] Gal I, Schlesinger O, Amir L, Alfonta L. Yeast surface display of dehydrogenases in microbial fuel-cells. Bioelectrochemistry. 2016;**112**:53-60

[53] Hubenova Y, Mitov M. Potential application of Candida melibiosica in biofuel cells. Bioelectrochemistry. 2010;**78**:57-61

[54] Babanova S, Hubenova Y, Mitov M. Influence of artificial mediators on yeast-based fuel cell performance. Journal of Bioscience and Bioengineering. 2011;**112**:379-387

[55] Hubenova Y, Rashkov R, Buchvarov V, Arnaudova M, Babanova S, Mitov M. Improvement of yeast-biofuel cell output by electrode modifications. Industrial and Engineering Chemistry Research. 2011;**50**:557-564

[56] Hubenova Y, Rashkov R, Buchvarov V, Babanova S, Mitov M. Nanomodified NiFe- and NiFeP-carbon felt as anode electrocatalysts in yeast-biofuel cell. Journal of Materials Science. 2011;**46**:7074-7081

[57] Shkil H, Schulte A, Guschin D, Schuhmann W. Electron transfer between genetically modified *Hansenula polymorpha* yeast cells and electrode surfaces via Os-complex modified redox polymers. ChemPhysChem. 2011;**12**:806-813

[58] Hasletta N, Rawsonb F, Barriërec F, Kunzed G, Pascoe N, Gooneratnea R, Baronian K. Characterisation of yeast microbial fuel cell with the yeast *Arxula adeninivorans* as the biocatalyst. Biosensors and Bioelectronics. 2011;**26**:3742-3747

[59] Kaneshiro H, Takano K, Takada Y, Wakisaka T, Tachibana T, Azuma M. A milliliter-scale yeast-based fuel cell with high performance. Biochemical Engineering Journal. 2014;**83**:90-96

[60] Ramanavicius A, Andriukonis E, Stirke A, Mikoliunaite L, Balevicius Z, Ramanaviciene A. Synthesis of polypyrrole within the cell wall of yeast byredox-cycling of $[Fe(CN)6]^{3-}$/ $[Fe(CN)6]^{4-}$. Enzyme and Microbial Technology. 2016;**83**:0-47

[61] Ostergaard S, Olsson L, Nielsen J. Metabolic engineering of *S. cerevisiae*. Microbiology and Molecular Biology Reviews. 2000;**64**:34-50

[62] Balint R, Cassidy N, Cartmell S. Conductive polymers: Towards a smart biomaterial for tissue engineering. Acta Biomaterialia. 2014;**10**:2341-2353

[63] Rahimnejad M, Ghoreyshi AA, Najafpour GD, Younesi H, Shakeri M. A novel microbial fuel cell stack for continuous production of clean energy. International Journal of Hydrogen energy. 2012;**37**:5992-6000

[64] Rabaey K, Boon N, Siciliano S, Verhaege M, Verstraete W. Biofuel cells select for microbial consortia that self-mediate electron transfer. Applied and Environmental Microbiology. 2004;**70**:5373-5382

Chapter 4

Advances in Metabolic Engineering of *Saccharomyces cerevisiae* for the Production of Industrially and Clinically Important Chemicals

Burcu Turanlı-Yıldız, Burcu Hacısalihoğlu and
Z. Petek Çakar

Additional information is available at the end of the chapter

http://dx.doi.org/10.5772/intechopen.70327

Abstract

Sustainable production of chemicals is of increasing importance, due to depletion of petroleum and environmental concerns. In addition to its importance in basic research as a simple, eukaryotic model organism, *Saccharomyces cerevisiae* has long been exploited in industry because of its physiological properties. And today, the development in genetic engineering toolbox and genome-scale metabolic models of *S. cerevisiae* has extended its application range to new products and bioprocesses. In addition, evolutionary engineering strategies have been useful in improving cellular properties of *S. cerevisiae*, such as tolerance to product toxicity and inhibitors. In this chapter, recent metabolic and evolutionary engineering studies that involve *S. cerevisiae* for the production of bulk chemicals and fine chemicals including flavours and pharmaceuticals are reviewed. It was shown that metabolic engineering particularly allowed the improvement of pharmaceuticals production, which will enable economic and large-scale production of many valuable pharmaceuticals. It is clear that *S. cerevisiae* will continue to be an important host for future metabolic engineering and metabolic pathway engineering applications to produce a variety of industrially and clinically important chemicals.

Keywords: pharmaceuticals, adaptive evolution, bulk chemicals, evolutionary engineering, flavours, fine chemicals, glutathione, metabolic engineering, organic acids, *Saccharomyces cerevisiae*

1. Introduction

Metabolic engineering was defined by Bailey [1] as 'the improvement of cellular activities by manipulation of enzymatic, transport and regulatory functions of the cell with the use of recombinant DNA technology'. More than 20 years after this first definition as a new scientific discipline, metabolic engineering has become an increasingly important research field of biotechnology. Today, metabolic engineering requires interdisciplinary work that includes molecular biology, applied microbiology, biochemical reaction engineering, biomedical research with the aid of high-throughput analytical tools in 'omics' research and bioinformatics [2].

There are two major approaches in metabolic engineering, as described by Bailey et al. [3], the rational metabolic engineering and inverse metabolic engineering. In rational metabolic engineering, extensive genetic and biochemical information is required on the metabolism or metabolic pathway of interest to make defined genetic manipulations. The cellular physiological responses are also complex. Thus, trying to re-engineer a cellular machine that is too complex and about which there is limited information is a major limitation in rational metabolic engineering. Difficulties in cloning in industrial strains due to the lack of relevant genetic tools, and GMO-concerns of the public regarding food industry are additional issues [2]. The inverse metabolic engineering approach was designed to avoid the above-mentioned limitations. Here, the desired phenotype is identified first, as a 'bottom-up' approach, and then its genetic and/or environmental basis is determined which is the most challenging step. However, owing to the powerful high-throughput analytical technologies in genomics, transcriptomics, proteomics and metabolomics, this step is becoming easier [2, 4]. Thus, without any need for extensive initial information on biochemistry, genetics and regulation on the organism of interest, the desired phenotype can be obtained. Adaptive evolution or evolutionary engineering, which is based on random mutation and selection by systematic cultivation of an initial microbial culture in the presence of a selective pressure to obtain desirable phenotypes [5], is a common inverse metabolic engineering strategy [2].

Metabolic engineering is a key strategy for harnessing microorganisms' ability to produce chemicals from renewable carbon sources. Microbial processes are attractive since they have significantly lower environmental impacts than the petroleum-based processes. However, the former is primarily an economic challenge. Therefore, it is vital to develop superior strains with improved yield, titer and productivity by engineering microbial physiology, stress response and metabolism [6]. Considering the market value of chemical products based on petroleum, the cost-competitive bio-based products, once achieved, would have significant economic value as replacements. It is estimated that the global market share of bio-based chemicals will rise from 2% in 2008 to 22% in 2025 [7].

In this chapter, we focused on the recent metabolic engineering studies that involve the baker's yeast, *Saccharomyces cerevisiae*, for the production of industrially and clinically important compounds. *S. cerevisiae* has many advantages to be used in metabolic engineering studies: it has 'Generally Recognized as Safe' (GRAS) status, and there is extensive information on its genetics, physiology and biochemistry. Besides being a common industrial microorganism for ethanol fermentation, baking, brewing, etc., *S. cerevisiae* has been regarded as a versatile cell factory for the production of a wide range of natural compounds by manipulation of the endogenous pathways and/or integration of heterologous pathways. In this review, metabolic engineering studies with *S. cerevisiae* are divided into two major categories: production of bulk chemicals, and production of fine chemicals including flavours and pharmaceuticals. Regarding the production of bulk chemicals, examples of organic acids that have potentials to be produced by fermentation at large-scale were discussed. As fine chemicals, glutathione and a variety of secondary metabolites used in food, cosmetic and health industries were discussed.

2. Production of bulk chemicals

The oil refinery is currently the major source of bulk chemicals such as solvents and polymer precursors. A significant portion of petroleum is used in the chemical catalysis for the production of chemicals and plastics [8]. However, in recent years, microbial production of chemicals based on renewable sources, such as biomass, has become important as a part of the efforts to reduce demand on diminishing petroleum and to reduce hazardous wastes. In addition, biotechnology makes new chemical monomers accessible, which are otherwise inaccessible due to high production cost [9].

In bio-refineries, the biomass is the first converted into simple sugars and then to valuable chemicals. Microorganisms are the main players of the latter conversion. Therefore, the development of a suitable strain for the particular process is needed. As a model yeast, *S. cerevisiae* has been a focus of metabolic engineering studies for the bio-based production of chemicals. 1,4-Diacids (succinic, fumaric and malic), itaconic acid, 3-hydroxypropionic acid and lactic acid are organic acids listed among the high-potential targets for industrial biotechnology [10]. Representative examples for the production of these bulk chemicals by metabolically engineered *S. cerevisiae* are summarized in **Table 1**.

Succinic acid is used in a wide range of industries from food to agriculture. Also, it has been considered as a generic intermediate for the bio-based polymers and can be a substitute of petroleum-derived maleic anhydride, which has a huge market [11]. Therefore, an increasing demand of succinic acid is expected in the future. Currently, it is mainly produced by chemical syntheses, which are based on petrochemical precursors. Biotechnological routes are pursued to achieve a sustainable production of succinic acid. *Anaerobiospirillum succiniciproducens* and *Actinobacillus succinogenes* are natural succinic acid producers. However, these organisms are prokaryotes that favour neutral pH for growth and require neutralization and a cost-additive product recovery process. In addition, there is a lack of suitable genetic tools for these organisms [12]. Although *S. cerevisiae* is not a natural producer of succinic acid as an end product, there have been efforts to metabolically engineer *S. cerevisiae*, since it has favourable properties such as the ability to operate at low pH values [13]. In general, the tricarboxylic acid (TCA) cycle and glyoxylate shunt are the focus of these studies. In order to redirect oxidative TCA pathway, elimination of succinate and isocitrate dehydrogenases has been proposed as a strategy. A yeast strain with disturbed TCA cycle due to deletions of *SDH1, SDH2, IDH1, IDP1*, produced succinic acid at a yield of 0.11 mol/mol glucose in shake

Bulk chemical produced	Representative studies and their strain improvement strategy [reference no]
Succinic acid	Disturbance of the citric acid cycle by deleting *SDH1, SDH2, IDH1, IDP1* [14] Disabled serine synthesis from glycolysis through a triple deletion of *SDH1, SER3* and *SER33* [15] Enhanced succinic acid export via heterologous expression of *MAE1* from *Schizosaccharomyces pombe* in *Saccharomyces cerevisiae SDH1*- and *SDH2*-disrupted strains [16]
Itaconic acid	Overexpression of *CAD* with a synthetic hybrid promoter and enhancement of flux towards the citric acid cycle by the sequential deletion of the *ADE3, BNA2* and *TES1* genes [19]
3-Hydroxypropionic acid	Reconstruction of malonyl-CoA to 3-HP pathway via expression of *MCR* from *Sulfolobus tokodaii* and *HPDH* from *Metallosphaera sedula* and increased precursor and cofactor availability [23] Reconstruction of β-alanine to 3-HP pathway via coexpression of *BAPAT* from *Bacillus cereus* and *HPDH* from *Escherichia coli* and redirection of flux towards β-alanine by overexpressing *PAND* from *Tribolium castaneum* [24] Reconstruction of malonyl-CoA to 3-HP pathway via coexpression of *MCR* from *Chloroflexus aurantiacus* and an inhibition-deficient *ACC1* and optimization of acetyl-CoA supply by overexpressing native *PDC1, ALD6, and ACS* from *Salmonella enterica* [25] Adaptive laboratory evolution for improved tolerance to 3-HP at pH 3.5 [45]
Lactic acid	Expression of genome-integrated *L-LDH* from bovine under *PDC1* promoter and inactivation of *PDC1* [28] Expression of genome-integrated *D-LDH* from *Leuconostoc mesenteroides* subsp. *mesenteroides* under *PDC1* promoter and inactivation of *PDC1* [29] Deletion of *PDC1* and expression of multiple copies of *L-LDH* from bovine [30] Inhibition of L-LDH consumption by deletion of *DLD1* and *JEN1*, elimination of ethanol and glycerol production by deleting *PDC1, ADH1, GPD1* and *GPD2*, and improvement of lactic acid tolerance by adaptive evolution and overexpression of *HAA1* [31] Overexpression of *HXT1* and *HXT7* hexose transporters [32] Repression of ethanol production by deleting *PDC1* and *ADH1* and enhanced acetyl-CoA supply by the introduction of the genes encoding acetylating acetaldehyde dehydrogenase enzyme from *Escherichia coli* [33] Enhancement of lactic acid transport by expressing *JEN1* and *ADY1* [34] Expression of *ESBP6*, a novel target isolated by screening a multi-copy yeast genomic DNA library [35]

Table 1. Bulk chemical production by metabolically engineered *S. cerevisiae*.

flask cultures [14]. A computational pathway prediction algorithm has been utilized to identify multiple gene deletion targets to redirect carbon fluxes towards succinic acid [15]. Three deletion targets, *SDH3*, *SER3* and *SER33*, were identified to couple succinic acid production to biomass formation. This strategy was based on the elimination of succinic acid consumption by the deletion of *SDH3* encoding cytochrome b subunit of succinate dehydrogenase. The serine biosynthesis was also disrupted by the deletions of *SER3* and *SER33*, which are paralogs encoding 3-phosphoglycerate dehydrogenase. Therefore, serine and glycine production were linked to succinic acid production via glyoxylate pathway. However, the engineered strain required glycine to be supplemented in the medium. Further, two successive laboratory evolution experiments for glycine prototrophy and faster growth were performed with this strain. Finally, overexpression of isocitrate lyase, Icl1p, in the evolved strain, resulted in a succinic acid yield of 0.07 mol/mol glucose under aerobic conditions without glycine addition. Metabolic profiling analysis of a succinic acid-producing recombinant *S. cerevisiae* hinted a metabolic engineering strategy involving expression of a malic acid transporter from *Schizosaccharomyces pombe* (*MAE1)* to export succinic acid out of cells [16].

Itaconic acid has currently application in the manufacture of pharmaceuticals, adhesives and resins. In addition, its polymerized form (polyitaconic acid) has potentials as a replacement of acrylic acid in the development of superabsorbents [17], and can be used in contact lenses, detergents and cleaners [18]. *Aspergillus terreus* is the present organism of choice for the industrial fermentation of itaconic acid. However, the process bears some constraints due to inherent characteristics of *A. terreus,* such as inhibition in the media and sensitivity to shear stress [19]. Kanamasa et al. isolated *cis*-aconitic acid decarboxylase (CAD), which is the key enzyme in the conversion of *cis*-aconitate to itaconic acid in *A. terreus,* and its heterologous expression in *S. cerevisiae* showed the possibility of itaconic acid production in yeast [20]. Blazeck et al. utilized a synthetic hybrid promoter carrying an enhancer and a core promoter module to optimize CAD expression in *S. cerevisiae* [19, 21]. A genome-wide metabolic model of the yeast was used to identify gene deletion targets to further increase the itaconic acid titer. Three sequential rounds of genome scan *in silico* highlighted three deletion targets; cytoplasmic trifunctional C1-tetrahydrofolate (THF) synthase, a putative tryptophan 2,3-dioxygenase or indoleamine 2,3-dioxygenase and a peroxisomal acyl-CoA thioesterase, encoded by *ADE3, BNA2* and *TES1,* respectively. The deletions rewired metabolic flux towards TCA cycle and enhanced itaconic acid titer (168 mg/L). However, further efforts are necessary to redirect carbon flux towards itaconic acid production in the yeast to approach titers obtained in *Aspergillus* species (>80 g/L).

3-Hydroxypropionic acid (3-HP) is another important platform chemical which can be produced from either sugars or glycerol and can be converted to 1,3-propanediol, acrylic acid, malonic acid, and acrylamide. 3-HP derivatives have a variety of applications in super absorbent polymers, surface coatings, adhesives and paints [11]. Although there are biological pathways to 3-HP via either glycerol, lactate, malonyl-CoA or β-alanine intermediates, no organism is known to produce it as an end product [22]. The pathways based on malonyl-CoA and β-alanine have been constructed in *S. cerevisiae* [23, 24]. Chen et al. evaluated different malonyl-CoA reductases. Malonyl-CoA reductase (MCR_{Ca}) from *Chloroflexus aurantiacus* was expressed in the yeast for the conversion of malonyl-CoA to 3-HP in a two-step reduction reaction. Further, carbon flux was redirected towards 3-HP through increasing the levels of malonyl-CoA and its immediate precursor, acetyl-CoA. For this purpose, native *ADH2* (alcohol dehydrogenase) and *ALD6* (NADP-dependent aldehyde dehydrogenase), and synthetic acs^{L641P}_{SE} (acetylation-insensitive acetyl-CoA synthetase from *Salmonella enterica*) were over-expressed to increase the level of acetyl-CoA. The cellular concentration of malonyl-CoA was increased by over-expression of *ACC1* (acetyl-CoA carboxylase), which is the sole enzyme in the conversion of acetyl-CoA to malonyl-CoA. Finally, 3-HP was produced at a titer of 463 mg/L when the production was coupled with enhanced supply of electron donor of MCR_{Ca} (NADPH) by heterologous expression of GAPNp (a non-phosphorylating glyceraldehyde-3-phosphate dehydrogenase from *Streptococcus mutans*) [23]. In another study, a significant improvement of 3-HP production was achieved when multiple copies of MCR were integrated into the yeast genome and a modified ACCp with phosphorylation deficiency was expressed. Finally, engineering of the redox metabolism of this strain produced 3-HP at a titer of 9.8 g/L in a glucose-limited, fed-batch system [25]. Borodina et al. utilized genome-scale modelling to compare the two biosynthetic routes in terms of maximum theoretical yields and identified β-alanine pathway as a more favourable route. They implemented the biosynthesis of 3-HP from glucose via β-alanine through coexpression of β-alanine-pyruvate aminotransferase from *Bacillus cereus* and 3-hydroxypropanoate dehydrogenase from *Escherichia coli*. Further, carbon-flux was redirected towards β-alanine by the supply of L-aspartate, the immediate precursor of β-alanine. The final strain yielded 3-HP at a titer of 13.7 g/L in glucose-limited fed-batch cultivation. In a similar fashion, production of 3-HP via both malonyl-CoA and β-alanine pathway was reported in a xylose-utilizing *S. cerevisiae* [24].

Lactic acid is a well-known fermentation product which is already widely used in food, cosmetics and pharmaceutical industries. Lactic acid derived from biomass is also valued as a monomer in the development of bioplastics [26]. Lactic acid bacteria, especially, *Lactobacillus* species, are often employed in lactic acid production. For large-scale lactic acid production, fermenting microorganisms with high acid tolerance, simple nutritional requirements and capability of growth at high cell density are pursued [27]. To this end, *S. cerevisiae* was engineered for lactic acid production by integrating lactate dehydrogenase (*LDH*) gene into its genome [28, 29]. Reduction in ethanol and glycerol production is desirable to direct metabolite fluxes to lactic acid production. Therefore, deletions of *PDH* encoding pyruvate dehydrogenase, *ADH* encoding alcohol dehydrogenase and *GPD1* encoding glycerol-3-phosphate dehydrogenase were reported to improve lactic acid production in *LDH*-expressing yeast strains [30, 31]. Another approach was the improvement of cell growth either by an increased glucose uptake via overexpression of hexose transporters (*HXT1* and *HXT7*) or an enhanced acetyl-CoA supply through implementing an acetyl-CoA synthesis pathway from *E. coli* in lactic acid-producing *S. cerevisiae* [32, 33]. In addition, elimination of NADH-consuming reactions through deletions of *NDE1* and *NDE2* encoding mitochondrial external NADH dehydrogenases was shown to improve lactic acid production due to increased cofactor availability. The yeast strains that expressed *JEN1* and *ADY2* encoding monocarboxylate permeases constitutively had improved lactic acid production due to higher efflux of lactic acid [34]. Recently, screening of a multi-copy genomic DNA library revealed a novel protein (ESBP6) involved in lactic acid adaptation response, although having a low similarity to monocarboxylate permeases [35]. Lactic acid accumulation under low pH conditions has detrimental effects on yeast cells. Therefore, tolerance to weak acids is another target to achieve high levels of organic acids like lactic acid. A recombinant *LDH*-expressing yeast strain was subjected to adaptive laboratory evolution in the presence of gradually increased D-lactic acid levels. A lactate over-producing strain was obtained with additional copies of *LDH* and *HAA1*, encoding a transcription activator involved in lactic acid stress, and a titer of 112 g/L was achieved in fed-batch cultivation [31].

Product toxicity is a major obstacle for achieving high titers of the target chemicals such as organic acids, aromatic substances and antibiotics [36]. There is limited knowledge about the molecular basis of the product toxicity and tolerance to enable a rational prediction of genetic changes [37]. David et al. developed, for the first time, a hierarchical dynamic pathway control system involving a two-stage fermentation concept and the use of a metabolic sensor in *S. cerevisiae* [38]. The growth and production phases were decoupled to allow sufficient biomass formation before accumulation of the product beyond toxic levels. In addition, they designed a metabolite sensor based on prokaryotic fapR-fapO system to regulate expression of pathway enzymes in relation to availability of metabolite pools during the production phase. Efficiency of this concept was demonstrated in 3-HP production, which was increased by 10-fold in titers. A more common, alternative approach used against product toxicity or toxic/inhibitory compounds is evolutionary engineering. It is particularly useful for obtaining genetically complex microbial phenotypes such as tolerance to inhibitors/toxic compounds or various stress types [39]. Successful results were obtained by our research group, regarding evolutionary engineering of multi-stress resistant [40], cobalt-resistant [41, 42], nickel-resistant [43], and ethanol-tolerant [44] *S. cerevisiae*. Another example for the use of evolutionary engineering against product toxicity involves adaptive evolution for lactic acid tolerance in *S. cerevisiae* [31]. Similarly, Kildegaard et al. isolated *S. cerevisiae* strains with resistance to 3-HP through laboratory evolution. Genome sequencing of the evolved strains and subsequent functional analyses identified a relevant mutation in *SFA1* gene (*S*-(hydroxymethyl) glutathione dehydrogenase) related to 3-HP tolerance [45].

3. Production of fine chemicals

Plant secondary metabolites hold the potential to be used as pharmaceuticals, cosmetic and food ingredients. However, the yield of these molecules when extracted from natural producers is not in sufficient amounts to meet industrial demands. In addition, chemical synthesis of these complex structures often requires multiple reaction steps and is not a commercially attractive route due to low product yields [46]. Currently, advances in metabolic engineering allowed commercial-scale microbial production of a number of fine chemicals [47–49]. Besides, there is an ongoing academic interest for reconstitution of biosynthetic pathways of several natural products, including complex pathways, in *S. cerevisiae*. Discovery of gene clusters involved in the biosynthesis of secondary metabolites have enhanced progress in microbial production of these molecules [50]. Computational studies have also been conducted to optimize heterologous production in a variety of industrial host microorganisms including *S. cerevisiae*, which involved application of flux balance analysis on genome-scale models for different hosts to identify the optimum host for production [51].

3.1. Flavours

Compounds belonging to isoprenoid and phenolics type of secondary metabolites are valued as natural fragrances and flavours. Flavour compounds can be produced from sugars (*de novo* synthesis) or from specific precursors (bioconversion) by using microorganisms.

Vanillin, a phenolic aldehyde, is one of the first flavour compounds produced in microbial hosts at commercial-scale. Current state of the microbial production of vanillin based on various precursors and the available production hosts have been recently reviewed by Gallage and Møller [52]. *De novo* biosynthesis of vanillin from glucose in *S. cerevisiae* has also been reported [53]. A multi-step conversion of a shikimate pathway intermediate (3-dehydroshikimate) to vanillin has been achieved through heterologous expression of four genes from *Podospora pauciseta, Nocardia iowensis, Corynebacterium glutamicum* and *Homo sapiens*. Once the vanillin biosynthesis was established, genome-scale metabolic modelling was used to identify gene deletion targets to improve vanillin production in *S. cerevisiae*. *PDC1* and *GDH1* deletions resulted in a five-fold increase in production (500 mg/L) [47].

p-Coumaric acid, a hydroxyl derivative of cinnamic acid, is a commercially attractive end-product and a platform compound for flavonoids, polyphenols and polyketides, as well. Rodriguez et al. achieved high titers (2 g/L) of *p*-coumaric acid as the end-product in *S. cerevisiae,* through optimization of native aromatic amino acid biosynthesis [54]. The competing pathways were eliminated while enhancing production pathways by the expression of feed-back resistant enzymes in combination with gene deletions and overexpression of analogue enzymes from *E. coli.*

β-Ionone is an apocarotenoid that is naturally present in raspberries. In *S. cerevisiae, de novo* synthesis of β-ionone was reported [55]. Beekwilder et al. constructed a β-carotene synthesis pathway via farnesyl diphosphate (FPP) intermediate through polycistronic expression of genes from *Xanthophyllomyces dendrorhous.* The pathway was further extended, for the first time, to produce β-ionone by the expression of a carotenoid-cleavage dioxygenase (*CCD1*) from raspberry.

2-Phenyl ethanol (2-PE) is another economically attractive flavour compound with a rose-like scent. Ehrlich pathway is involved in the bioconversion of phenylalanine to 2-phenyl ethanol within *S. cerevisiae*. Elimination of allosteric feedback regulation on the aromatic amino acid biosynthesis resulted in an increase of up to 200-fold in the production of aromatic compounds, including 2-PE. Romagnoli et al. constructed a deletion library of non-essential genes in *S. cerevisiae* by Synthetic Genetic Array (SGA) technology and identified that *ARO8* encoding an aromatic amino acid transaminase is a target to improve phenylethanol production from glucose [56]. Recently, Shen et al. identified *AAT2* encoding a cytosolic aspartate aminotransferase as another deletion target [57]. Deletion of these two genes in combination with the overexpression of Ehrlich pathway enzymes resulted in a significant improvement in 2-PE production from glucose, at a titer of 96 mg/L.

3.2. Pharmaceuticals

Another major area of metabolic engineering research is the production of clinically important compounds. In this section, examples will be given for the production of a variety of such compounds by metabolically engineered yeast. Representative examples for the production of pharmaceuticals by metabolically engineered *S. cerevisiae* are summarized in **Table 2**.

Glutathione, a naturally occurring tripeptide, is an important compound used in health and cosmetic industries. It is produced by using *S. cerevisiae* at commercial-scale. There has been a remarkable progress in glutathione production by metabolic engineering studies over the last few decades. Improved levels of glutathione production were achieved by *YAP1*

Pharmaceutical produced	Representative studies and their strain improvement strategy [reference no]
Glutathione	Overexpression of *YAP1* [58] Manipulation of the sulphate assimilation pathway by overexpressing *MET14* and *MET16* [59] Improved oxidized glutathione production by overexpression of *GSH1, GSH2,* and *ERV1* and the deletion of *GLR1* [60] Adaptive laboratory evolution in the presence of increasing levels of acrolein and screening for enhanced glutathione production [61] Whole-genome engineering via genome shuffling and screening for enhanced glutathione production [62]
Artemisinin/artemisinic acid	Reconstruction of the complete biosynthetic pathway of artemisinic acid, including the three-step oxidation of amorphadiene to artemisinic acid by expression of *CYP71AV1, CPR1, CYB5, ADH1* and *ALDH1* from *Artemisia annua* [48]
Taxol/taxadiene	Expression of a truncated version of the endogenous *tHMG1* and *GGPPS* from *Taxus chinensis* or *Sulfolobus acidocaldarius* together with *TDC1* from *T. chinensis* [66] Prediction of the efficiency of different GGPPS enzymes via computer aided protein modelling [67]
Forskolin	Expression of a promiscuous cytochrome P450 from *Salvia pomifera* [68]
Polyketides	Heterologous expression of 6-MSA synthase gene from *Penicillium patulum* together with PPTases from either *Bacillus subtilis* or *Aspergillus nidulans* [69] Construction of polyketide precursor pathways by expressing *prpE* from *Salmonella typhimurium* and PCC pathway from *Streptomyces coelicolor* [70] Enhanced cofactor supply by expressing *2-PS* from *Gerbera hybrida* [71]
Resveratrol	Reconstruction of a *de novo* pathway by expressing *TAL* from *Herpetosiphon aurantiacus, 4-CL1* from *Arabidopsis thaliana* and *VST1* from *Vitis vinifera* [49] Expression of *4CL1* from *A. thaliana* and *STS* from *Arachis hypogaea* [73] Expression of *PAL* from *Rhodosporidium toruloides, C4H* and *4-CL1* from *A. thaliana,* and *STS* from *A. hypogaea* [74] Expression of 4-coumaroyl-coenzyme A ligase (*4CL1*) from *A. thaliana* and stilbene synthase (*STS*) from *V. vinifera* [75] Overexpression of the resveratrol biosynthesis pathway, enhancement of P450 activity, increasing the precursor supply for resveratrol synthesis via phenylalanine pathway [76]
Dihydrochalcones	Expression of the heterologous pathway genes in a *TSC13*-overexpressing *S. cerevisiae* strain [78]
Alkaloids	Expression of 14 monoterpene indole alkaloid pathway genes from *Catharanthus roseus* and enhanced secondary metabolism to produce strictosidine *de novo* [79] Construction of the complete *de novo* biosynthetic pathway to norcoclaurine by expressing a mammalian TyrH enzyme and DODC from *Pseudomonas putida,* along with four genes required for biosynthesis of its electron carrier cosubstrate [80] Expression of AdoMet-dependent methyltransferase enzymes (6-OMT, CNMT and 4'-OMT) from plant and human origin to produce reticuline from norlaudanosoline [81] Reconstruction of berberine biosynthetic pathway from reticuline by expressing seven relevant heterologous genes [82] Reconstruction of a 10-gene biosynthetic pathway from plant to produce sanguinarine from norlaudanosoline [83] Expression of 16 heterologous plant enzymes to produce noscapine from canadine [84] Reconstruction of a seven-gene pathway for the production of codeine and morphine from (*R*)-reticuline [85] Reconstruction of a *de novo* biosynthetic pathway for thebaine by expression of 21 genes from plants, mammals, bacteria and the yeast [86]

Table 2. Production of pharmaceuticals by metabolically engineered *S. cerevisiae*.

overexpression [58], metabolic engineering of the yeast sulphate assimilation pathway and glutathione biosynthetic pathway [59], overexpression of a novel glutathione export ABC protein (Adp1p, Gxa1p) and the engineered thiol redox metabolism [60]. Also, the inverse metabolic engineering approach was used to increase glutathione production in *S. cerevisiae* [61, 62]. In an evolutionary engineering study, acrolein, a toxic α,β-unsaturated aldehyde, was used as a selection agent. Two rounds of adaptive evolution in the presence of increasing levels of acrolein resulted in evolved strains with acrolein tolerance and up-to 3.3-fold higher glutathione accumulation in comparison to the parental strain [61]. Genome shuffling has also been applied to obtain yeast strains with increased glutathione content. Two rounds of recursive protoplast fusion were performed with the improved strains initially obtained from ultraviolet irradiation and chemical mutagenesis. The strain with highest glutathione content showed 9.9-fold transcriptional up-regulation of glutathione synthetase gene (*GSH-I*) [62].

Terpene derivatives are economically viable molecules that are used in the synthesis of drugs such as the antimalarial agent artemisinin, and the anticancer agent taxol [63]. Several terpenoids have been produced in *S. cerevisiae* by reconstitution of the relevant biosynthetic pathways. As part of efforts to establish a solid source of artemisinin, *S. cerevisiae* was metabolically engineered to produce artemisinic acid, which is an artemisinin precursor [48]. As the microbially produced artemisinic acid was converted to artemisinin by synthetic chemistry methods, that study was reported as a good example for combining biological production by metabolic engineering with production by synthetic chemistry [64]. Paddon et al. have, for the first time, designed a *S. cerevisiae* strain with the complete biosynthetic pathway of artemisinic acid, involving overexpression of the mevalonate pathway enzymes, and achieved commercial-scale titers (25 g/L) [48].

The well-known diterpenoid taxol is an anti-cancer agent [63, 65]. As a first step towards taxol production, *S. cerevisiae* was metabolically engineered for taxadiene biosynthesis [66]. For this purpose, heterologous genes encoding enzymes from the early steps of the taxoid biosynthesis pathway, isoprenoid pathway, were introduced, along with a regulatory factor to inhibit competing pathways. The results were promising enough for taxol production in recombinant microorganisms [66]. By using protein modelling and substrate docking, different geranylgeranyl diphosphate synthases were screened and expressed in a recombinant taxadiene-producing yeast. The yeast strains were compared in terms of their metabolism using metabolomics approach to identify an efficient host for taxadiene production [67].

Forskolin is a labdene diterpene with potentials to be used in the treatment of blood pressure, in weight-loss supplements and in the protection against congestive heart failure. Ignea et al. constructed a yeast platform to produce 11β-hydroxy-manoyl oxide, forskolin precursor. Although the forskolin biosynthetic pathway has not been completely discovered yet, a promiscuous cytochrome P450 from *Salvia pomifera* was identified as a replacement to achieve the synthesis of the forskolin precursor. This study can provide a basis for the biosynthesis of various tricyclic (8,13)-epoxy-labdanes [68].

Polyketides are also a major group of natural products with a wide range of applications as antibiotics, immunosuppressors, cholesterol lowering agents and other drugs [69]. *S. cerevisiae* is known as a suitable production host for simple polyketides. An earlier study demonstrated the production of a simple polyketide, 6-methylsalicylic acid, by heterologous expression of 6-methylsalicylic acid synthase in *S. cerevisiae* [69]. However, the major challenge in the synthesis of complex polyketides was the lack of polyketide precursor pathways in *S. cerevisiae.* To overcome this, a relevant pathway was introduced into *S. cerevisiae* to produce a precursor for complex polyketides, methylmalonyl-coenzyme A (CoA). This engineered yeast strain had the capability of the production of a triketide lactone, when supplemented with propyldiketide thioester [70]. Since polyketides are derived from acetyl-CoA and malonyl-CoA precursors, an increase in the acetyl-CoA and the cofactor (NADPH) in a yeast strain expressing 2-pyrone synthase (2-PS) from *Gerbera hybrida* led to 6.4-fold higher triacetic acid lactone production, compared to the reference strain [71].

The strategy of engineered precursor pools has also been applied in the production of resveratrol. Resveratrol is a polyketide derivative with potent antioxidant properties and it has been recently brought to market as a bio-product [72]. Earlier reports on the production of resveratrol were based on bioconversion of aromatic precursors such as *p*-coumaric acid and tyrosine by engineered *S. cerevisiae* strains [73, 74]. The highest resveratrol titer achieved by using this approach was obtained by an engineered industrial Brazilian *S. cerevisiae* strain, at a titer of 391 mg/L resveratrol on complex medium supplemented with *p*-coumaric acid [75]. Recently, in order to produce resveratrol from cheaper carbon sources, *de novo* biosynthesis of resveratrol via tyrosine intermediate in *S. cerevisiae* has been established by constructing an engineered pathway, involving tyrosine ammonia-lyase from *Herpetosiphon aurantiacus,* 4-coumaryl-CoA ligase from *Arabidopsis thaliana* and resveratrol synthase from *Vitis vinifera* [49]. To direct flux towards tyrosine, feedback-insensitive *ARO4* encoding 3-deoxy-D-arabino-heptulosonate-7-phosphate synthase and *ARO7* encoding a chorismate mutase were overexpressed. To increase the precursor malonyl-CoA, an inactivation-sensitive acetyl-CoA carboxylase was overexpressed. Resveratrol production was further improved by integration of multiple copies of pathway genes, and finally, a titer of 415.65 and 531.41 mg/L resveratrol was obtained in a fed-batch cultivation with glucose or ethanol as the carbon source, respectively [76]. Koopman et al. also focused on deregulation of feedback mechanism of aromatic amino acid biosynthesis for *de novo* production of naringenin, which is an important platform molecule for the production of flavonoids [77].

Dihydrochalcones (DHCs) such as nothofagin, phlorizin and naringin dihydrochalcone are another group of polyketide derivatives with commercial value as antioxidants, antidiabetics or sweeteners. Recently, *de novo* synthesis of DHCs via phloretin intermediate has been reported in *S. cerevisiae* [78]. First, phloretin biosynthesis was achieved with the aid of a side activity of an endogenous double-bond reductase, in combination with heterologous pathway enzymes. To eliminate by-product formation, a chalcone synthase with high substrate specificity was expressed from *Hordeum vulgare*. Commencing with phloretin, several DHC derivatives with antioxidant, antidiabetic and sweetener properties have been obtained through an extension pathway involving methylation or glycosylation by previously known enzymes.

Recently, there have also been many reports on the reconstitution of biosynthetic pathways of alkaloids in yeast. Alkaloids are nitrogen-containing complex molecules with potent biological activity. Currently, there are around 50 alkaloid-based drugs, including the anticancer drug vincristine, the antitussive agent noscapine and the analgesic codeine. Strictosidine was the first reported plant-derived alkaloid produced *de novo* in *S. cerevisiae* [79]. Strictosidine is a common intermediate of a list of alkaloids derived from tryptophan in plants, including the antimalarial quinine and anticancer agent vincristine [79]. Brown et al. reconstituted its biosynthetic pathway in *S. cerevisiae.* To enable strictosidine production in yeast, 14 genes from *Catharanthus roseus* were expressed [79]. The flux through the pathway was further improved by integration of additional copies of the relevant endogenous genes and three gene deletions that eliminated competing pathways. *S. cerevisiae* has also been engineered for the production of (*S*)-reticuline, which is a key branch point intermediate in the biosynthesis of a variety of alkaloids, including well-known opioids such as morphine and thebaine [80]. Bioconversion of a commercial substrate norlaudanosoline to reticuline was reported in an engineered yeast strain expressing three different AdoMet-dependent methyltransferase enzymes (6-OMT, CNMT and 4′-OMT) from plant and human origin [81]. Trenchard et al. constructed a route to reticuline which enabled *de novo* synthesis of this molecule via norcoclaurine intermediate, the actual intermediate in plants. The pathway comprised of a modified yeast amino acid biosynthesis pathway, in combination with a heterologous pathway involving seven relevant enzymes [80]. In other studies, *S. cerevisiae* strains were engineered to produce berberine, dihydrosanguinarine and noscapine from norlaudanosoline via reticuline intermediate, through a 7-, 10- and 14-step pathway involving heterologous expression of plant enzymes, respectively [82–84]. Also, the production of codeine and morphine from (*R*)-reticuline was reported by reconstitution of a seven-gene pathway in *S. cerevisiae* [85]. These studies provided a basis towards designing yeast cell factories for *de novo* production of reticuline-derived molecules. Recently, a complete pathway of biosynthesis of opioid thebaine from sugar has been established in *S. cerevisiae* [86]. This work involved a combination of enzyme discovery, protein engineering of a key cytochrome P450 and pathway optimization. The thebaine-producing yeast strains required expression of 21 heterologous genes from plants, mammals, bacteria and yeast. The pathway was also extended through expression of two additional genes from bacteria and plant to produce hydrocodone, a widely prescribed opioid drug.

4. Summary and outlook

For fine chemicals such as amino acids, vitamins, flavours, nutraceuticals, organic acids and fragrances, profit margins are usually not high and could be affected by substrate availability and cost. However, metabolic engineering enabled improvements in production of both pharmaceuticals and fine chemicals which will allow economic and large-scale production of many valuable compounds in near future.

It is obvious that *S. cerevisiae* will continue to be an important host for future metabolic engineering applications. There will be more comprehensive future studies on the production of chemicals by metabolic engineering of *S. cerevisiae.* These metabolic engineering strategies will most likely involve combinations of rational and inverse metabolic engineering approaches by adaptive evolution of recombinant *S. cerevisiae* with engineered metabolic pathways for various substrate utilization. Additionally, more studies on adaptive evolution and molecular characterization of tolerance to toxic end-products are expected in the future. Similarly, metabolic pathway engineering of *S. cerevisiae* will allow efficient production of more clinically important compounds and fine chemicals. It can be predicted that the advances in systems biology and bioinformatics will make a significant contribution to yeast metabolic engineering.

Acknowledgements

We thank Arman Akşit for technical assistance. Our research presented in this review was supported by the Scientific and Technological Research Council of Turkey (TUBITAK) (project nos: 105T314 and 107T284, PI: ZPÇ), TUBITAK-COST (project no: 109T638, PI: ZPÇ, COST Action no: CM0902), Federation of European Microbiological Societies (FEMS) Research Fellowship (2009-2 to BTY), Turkish State Planning Organization (DPT) and Istanbul Technical University Research Funds (project nos: 30108, 33237, 34200; PI: ZPÇ). BH is financially supported by the Faculty Member Training Programme (ÖYP) of the Council of Higher Education (YÖK), Turkey. We also would like to thank former graduate students Urartu Ö.Ş. Şeker, Ceren Alkım, Tuğba Sezgin, Ülkü Yılmaz, Gökhan Küçükgöze and Berrak Gülçin Balaban, as well as our colleagues and collaborators Uwe Sauer, Jean Marie François, Laurent Benbadis, Süleyman Akman, Bülent Balta, Candan Tamerler and Mehmet Sarıkaya for their scientific contribution in our research presented in this review.

Author details

Burcu Turanlı-Yıldız[1,2,†], Burcu Hacısalihoğlu[1,2,†] and Z. Petek Çakar[1,2*]

*Address all correspondence to: cakarp@itu.edu.tr

1 Department of Molecular Biology and Genetics, Faculty of Science and Letters, Istanbul Technical University, Maslak, Istanbul, Turkey

2 Istanbul Technical University, Dr. Orhan Öcalgiray Molecular Biology, Biotechnology and Genetics Research Center (ITU-MOBGAM), Maslak, Istanbul, Turkey

† These authors have equal contribution to this work

References

[1] Bailey JE. Toward a science of metabolic engineering. Science. 1991;**252**:1668-1675

[2] Cakar ZP, Turanli-Yildiz B, Alkim C, Yilmaz U. Evolutionary engineering of *Saccharomyces cerevisiae* for improved industrially important properties. FEMS Yeast Research. 2012;**12**:171-182

[3] Bailey JE, Sburlati A, Hatzimanikatis V, Lee K, Renner WA, Tsai PS. Inverse metabolic engineering: A strategy for directed genetic engineering of useful phenotypes. Biotechnology and Bioengineering. 1996;**52**:109-121

[4] Bro C, Nielsen J. Impact of "ome" analyses on inverse metabolic engineering. Metabolic Engineering. 2004;**6**:204-211

[5] Winkler JD, Kao KC. Recent advances in the evolutionary engineering of industrial biocatalysts. Genomics. 2014;**104**:406-411

[6] Chubukov V, Mukhopadhyay A, Petzold CJ, Keasling JD, Martín HC. Synthetic and systems biology for microbial production of commodity chemicals. NPJ Systems Biology and Applications. 2016;**2**:16009

[7] Biddy MJ, Scarlata C, Kinchin C. Chemicals from Biomass: A Market Assessment of Bioproducts with Near-Term Potential. No. NREL/TP-5100-65509. Golden, CO: National Renewable Energy Laboratory (NREL); 2016

[8] Adkins J, Pugh S, McKenna R, Nielsen DR. Engineering microbial chemical factories to produce renewable "biomonomers." Frontiers in Microbiology. 2012;**3**:1-12

[9] Cherubini F. The biorefinery concept : Using biomass instead of oil for producing energy and chemicals. Energy Conversation and Management. 2010;**51**:1412-1421

[10] Bozell JJ, Peterson GR. Technology development for the production of biobased products from biorefinery carbohydrates — the US Department of Energy's "Top 10" revisited. Green Chemistry. 2010;**12**:539-554

[11] Patel M, Crank M, Dornburg V, Hermann B, Roes L, Husing B, Overbeek L, Terragni F, Recchia E. Medium and long-term opportunities and risks of the biotechnological production of bulk chemicals from renewable resources: The Potential of White Biotechnology. Utrecht, Netherlands: The BREW Project; 2006

[12] Beauprez JJ, De Mey M, Soetaert WK. Microbial succinic acid production : Natural versus metabolic engineered producers. Process Biochemistry. 2010;**45**:1103-1114

[13] Ahn JH, Jang Y, Lee SY. Production of succinic acid by metabolically engineered microorganisms. Current Opinion in Biotechnology. 2016;**42**:54-66

[14] Raab AM, Gebhardt G, Bolotina N, Weuster-Botz D, Lang C. Metabolic engineering of *Saccharomyces cerevisiae* for the biotechnological production of succinic acid. Metabolic Engineering. 2010;**12**:518-525

[15] Agren R, Otero JM, Nielsen J. Genome-scale modeling enables metabolic engineering of *Saccharomyces cerevisiae* for succinic acid production. Journal of Industrial Microbiology and Biotechnology. 2013;**40**:735-747

[16] Ito Y, Hirasawa T, Shimizu H. Metabolic engineering of *Saccharomyces cerevisiae* to improve succinic acid production based on metabolic profiling. Bioscience Biotechnology and Biochemistry. 2014;**78**:151-159

[17] Shi D, Gao Y, Sun L, Chen M. Superabsorbent poly(acrylamide co itaconic acid) hydrogel microspheres : Preparation, characterization and absorbency. Polymer Science Series A. 2014;**56**:275-282

[18] Willke T, Vorlop K-D. Biotechnological production of itaconic acid. Applied Microbiology and Biotechnology. 2001;**56**:289-295

[19] Blazeck J, Miller J, Pan A, Gengler J, Holden C, Jamoussi M, Alper HS. Metabolic engineering of *Saccharomyces cerevisiae* for itaconic acid production. Applied Microbiology and Biotechnology. 2014;**98**:8155-8164

[20] Kanamasa S, Dwiarti L, Okabe M, Park EY. Cloning and functional characterization of the cis-aconitic acid decarboxylase (CAD) gene from *Aspergillus terreus*. Applied Microbiology and Biotechnology. 2008;**80**:223-229

[21] Blazeck J, Garg R, Reed B, et al. Controlling promoter strength and regulation in *Saccharomyces cerevisiae* using synthetic hybrid promoters. Biotechnology and Bioengineering. 2012;**109**:2884-2895

[22] Jiang X, Meng X, Xian M. Biosynthetic pathways for 3-hydroxypropionic acid production. Applied Microbiology and Biotechnology. 2009;**82**:995-1003

[23] Chen Y, Bao J, Kim I, Siewers V, Nielsen J. Coupled incremental precursor and co-factor supply improves 3-hydroxypropionic acid production in *Saccharomyces cerevisiae*. Metabolic Engineering. 2014;**22**:104-109

[24] Borodina I, Kildegaard KR, Jensen NB, Blicher TH, Maury J, Sherstyk S, Schneider K, Lamosa P, Herrgård MJ, Rosenstand I, Öberg F, Forster J, Nielsen J. Establishing a synthetic pathway for high-level production of 3-hydroxypropionic acid in *Saccharomyces cerevisiae* via β-alanine. Metabolic Engineering. 2015;**27**:57-64

[25] Kildegaard KR, Jensen NB, Schneider K, Czarnotta E, Özdemir E, Klein T, Maury J, Ebert BE, Christensen HB, Chen Y, Kim IK, Herrgård MJ, Blank LM, Forster J, Nielsen J, Borodina I. Engineering and systems-level analysis of *Saccharomyces cerevisiae* for production of 3-hydroxypropionic acid via malonyl-CoA reductase-dependent pathway. Microbial Cell Factories. 2016;**15**:53

[26] Lunt J. Large-scale production, properties and commercial applications of polylactic acid polymers. Polymer Degradation and Stability. 1998;**59**:145-152

[27] Eiteman MA, Ramalingam S. Microbial production of lactic acid. Biotechnology Letters. 2015;**37**:955-972

[28] Ishida N, Saitoh S, Tokuhiro K, Nagamori E, Matsuyama T, Kitamoto K, Takahashi H. Efficient production of L-lactic acid by metabolically engineered *Saccharomyces cerevisiae* with a genome-integrated L-lactate dehydrogenase gene. Applied and Environmental Microbiology. 2005;**71**:1964-1970

[29] Ishida N, Suzuki T, Tokuhiro K, et al. D-Lactic acid production by metabolically engineered *Saccharomyces cerevisiae*. Journal of Bioscience and Bioengineering. 2006;**101**: 172-177

[30] Ishida N, Saitoh S, Ohnishi T, Tokuhiro K, Nagamori E, Kitamoto K, Takahashi H. Metabolic engineering of *Saccharomyces cerevisiae* for efficient production of pure L-(+)-lactic acid. In: McMillan JD, Adney WS, Mielenz JR, Klasson KT, editors. Twenty-Seventh Symposium on Biotechnology for Fuels and Chemicals. Totowa, New Jersey: Humana Press; 2006. pp. 795-807

[31] Baek S, Kwon EY, Kim YH, Hahn J. Metabolic engineering and adaptive evolution for efficient production of D-lactic acid in *Saccharomyces cerevisiae*. Applied Microbiology and Biotechnology. 2016;**100**:2737-2748

[32] Rossi G, Sauer M, Porro D, Branduardi P. Effect of HXT1 and HXT7 hexose transporter overexpression on wild-type and lactic acid producing *Saccharomyces cerevisiae* cells. Microbial Cell Factories. 2010;**9**:15

[33] Song J, Park J, Kang CD, Cho H, Yang D, Lee S, Myung K. Introduction of a bacterial acetyl-CoA synthesis pathway improves lactic acid production in *Saccharomyces cerevisiae*. Metabolic Engineering. 2015;**35**:38-45

[34] Pacheco A, Sa-Pessoa J, Bessa D, Goncalves MJ, Paiva S, Casal M, Queiros O. Lactic acid production in *Saccharomyces cerevisiae* is modulated by expression of the monocarboxylate transporters Jen1 and Ady2. FEMS Yeast Research. 2012;**12**:375-381

[35] Sugiyama M, Akase S, Nakanishi R, Kaneko Y, Harashima S. Overexpression of ESBP6 improves lactic acid resistance and production in *Saccharomyces cerevisiae*. Journal of Bioscience and Bioengineering. 2016;**122**:415-420

[36] Skyta B. Techniques in Applied Microbiology. Prague: Elsevier; 1995. p. 122

[37] Dragosits M, Mattanovich D. Adaptive laboratory evolution—Principles and applications for biotechnology. Microbial Cell Factories. 2013;**12**:64

[38] David F, Nielsen J, Siewers V. Flux control at the malonyl-CoA node through hierarchical dynamic pathway regulation in *Saccharomyces cerevisiae*. ACS Synthetic Biology. 2016;**5**:224-233

[39] Alkım C, Turanlı-Yıldız B, Çakar ZP. Evolutionary engineering of yeast. In: Mapelli V, editor. Yeast Metabolic Engineering: Methods and Protocols. New York: Springer; 2014. pp. 169-183

[40] Cakar ZP, Seker UOS, Tamerler C, Sonderegger M, Sauer U. Evolutionary engineering of multiple-stress resistant *Saccharomyces cerevisiae*. FEMS Yeast Research. 2005;**5**:569-578

[41] Cakar ZP, Alkim C, Turanli B, Tokman N, Akman S, Sarikaya M, Tamerler C, Benbadis L, François JM. Isolation of cobalt hyper-resistant mutants of *Saccharomyces cerevisiae* by *in vivo* evolutionary engineering approach. Journal of Biotechnology. 2009;**143**:130-138

[42] Alkim C, Benbadis L, Yilmaz U, Cakar ZP, Francois JM. Mechanisms other than activation of the iron regulon account for the hyper-resistance to cobalt of a *Saccharomyces cerevisiae* strain obtained by evolutionary engineering. Metallomics. 2013;**5**:1043-1060

[43] Küçükgöze G, Alkım C, Yılmaz Ü, Kısakesen Hİ, Gündüz S, Akman S, Çakar ZP. Evolutionary engineering and transcriptomic analysis of nickel-resistant *Saccharomyces cerevisiae*. FEMS Yeast Research. 2013;**13**:731-734

[44] Turanlı-Yıldız B, Benbadis L, Alkım C, Sezgin T, Akşit A, Gökçe A, Öztürk Y, Baykal AT, Çakar ZP, François JM. *In vivo* evolutionary engineering for ethanol-tolerance of *Saccharomyces cerevisiae* haploid cells triggers diploidization. Journal of Bioscience and Bioengineering. 2017;**124**: DOI: 10.1016/j.jbiosc.2017.04.012

[45] Kildegaard KR, Hallström BM, Blicher TH, Sonnenschein N, Jensen NB, Sherstyk S, Harrison SJ, Maury J, Herrgård MJ, Juncker AS, Forster J, Nielsen J, Borodina I. Evolution reveals a glutathione-dependent mechanism of 3-hydroxypropionic acid tolerance. Metabolic Engineering. 2014;**26**:57-66

[46] Siddiqui MS, Thodey K, Trenchard I, Smolke CD. Advancing secondary metabolite biosynthesis in yeast with synthetic biology tools. FEMS Yeast Research. 2012;**12**:144-170

[47] Brochado AR, Matos C, Møller BL, Hansen J, Mortensen UH, Patil KR. Improved vanillin production in baker's yeast through *in silico* design. Microbial Cell Factories. 2010;**9**:84

[48] Paddon CJ, Westfall PJ, Pitera DJ, Benjamin K, Fisher K, McPhee D, et al. High-level semi-synthetic production of the potent antimalarial artemisinin. Nature. 2013;**496**:528-532

[49] Li M, Kildegaard KR, Chen Y, Rodriguez A, Borodina I, Nielsen J. *De novo* production of resveratrol from glucose or ethanol by engineered *Saccharomyces cerevisiae*. Metabolic Engineering. 2015;**32**:1-11

[50] Billingsley JM, DeNicola AB, Tang Y. Technology development for natural product biosynthesis in *Saccharomyces cerevisiae*. Current Opinion in Biotechnology. 2016;**42**:74-83

[51] Boghigian BA, Lee K, Pfeifer BA. Computational analysis of phenotypic space in heterologous polyketide biosynthesis-applications to *Escherichia coli, Bacillus subtilis,* and *Saccharomyces cerevisiae*. Journal of Theoretical Biology. 2010;**262**:197-207

[52] Gallage NJ, Møller BL. Vanillin—Bioconversion and bioengineering of the most popular plant flavor and its *de novo* biosynthesis in the vanilla orchid. Molecular Plant. 2015; **8**:40-57

[53] Hansen EH, Møller BL, Kock GR, Bu CM, Kristensen C, Jensen OR, Okkels FT, Olsen CE, Motawia MS, Hansen J. *De novo* biosynthesis of vanillin in fission yeast (*Schizosaccharomyces pombe*) and baker's yeast (*Saccharomyces cerevisiae*). Applied and Environmental Microbiology. 2009;**75**:2765-2774

[54] Rodriguez A, Kildegaard KR, Li M, Borodina I, Nielsen J. Establishment of a yeast platform strain for production of p-coumaric acid through metabolic engineering of aromatic amino acid biosynthesis. Metabolic Engineering. 2015;**31**:181-188

[55] Beekwilder J, van Rossum HM, Koopman F, Sonntag F, Buchhaupt M, Schrader J, Hall RD, Bosch D, Pronk JT, van Maris AJA, Daran J. Polycistronic expression of a β-carotene

biosynthetic pathway in *Saccharomyces cerevisiae* coupled to β-ionone production. Journal of Biotechnology. 2014;**192**:383-392

[56] Romagnoli G, Knijnenburg TA, Liti G, Louis EJ, Pronk JT, Daran J. Deletion of the *Saccharomyces cerevisiae* ARO8 gene, encoding an aromatic amino acid transaminase, enhances phenylethanol production from glucose. Yeast. 2015;**32**:29-45

[57] Shen L, Nishimura Y, Matsuda F, Ishii J, Kondo A. Overexpressing enzymes of the Ehrlich pathway and deleting genes of the competing pathway in *Saccharomyces cerevisiae* for increasing 2-phenylethanol production from glucose. Journal of Bioscience and Bioengineering. 2016;**122**:34-39

[58] Orumets K, Kevvai K, Nisamedtinov I, Tamm T, Paalme T. *YAP1* over-expression in *Saccharomyces cerevisiae* enhances glutathione accumulation at its biosynthesis and substrate availability levels. Biotechnology Journal. 2012;**7**:566-568

[59] Hara KY, Kiriyama K, Inagaki A. Improvement of glutathione production by metabolic engineering the sulfate assimilation pathway of *Saccharomyces cerevisiae*. Applied Microbiology and Biotechnology. 2012;**94**:1313-1319

[60] Hara KY, Aoki N, Kobayashi J, Kiriyama K. Improvement of oxidized glutathione fermentation by thiol redox metabolism engineering in *Saccharomyces cerevisiae*. Applied Microbiology and Biotechnology. 2015;**99**:9771-9778

[61] Anett P, Steiger MG, Caterina H, Lang C, Mattanovich D, Sauer M. Enhanced glutathione production by evolutionary engineering of *Saccharomyces cerevisiae* strains. Biotechnology Journal. 2015;**10**:1719-1726

[62] Yin H, Ma Y, Deng Y, Xu Z, Liu J, Zhao J, Dong J. Genome shuffling of *Saccharomyces cerevisiae* for enhanced glutathione yield and relative gene expression analysis using fluorescent quantitation reverse transcription polymerase chain reaction. Journal of Microbiological Methods. 2016;**127**:188-192

[63] Ye VM, Bhatia SK. Metabolic engineering for the production of clinically important molecules: Omega-3 fatty acids, artemisinin, and taxol. Biotechnology Journal. 2012;**7**:20-33

[64] Keasling JD. Manufacturing molecules through metabolic engineering. Science. 2010;**330**: 1355-1358

[65] Pscheidt B, Glieder A. Yeast cell factories for fine chemical and API production. Microbial Cell Factories. 2008;**7**:25

[66] Engels B, Dahm P, Jennewein S. Metabolic engineering of taxadiene biosynthesis in yeast as a first step towards Taxol (Paclitaxel) production. Metabolic Engineering. 2008; **10**:201-206

[67] Ding MZ, Yan HF, Li LF, Zhai F, Shang LQ, Yin Z, Yuan YJ. Biosynthesis of taxadiene in *Saccharomyces cerevisiae*: Selection of geranylgeranyl diphosphate synthase directed by a computer-aided docking strategy. PLoS One. 2014;**9**:10 e109348

[68] Ignea C, Ioannou E, Georgantea P, Trikka FA, Athanasakoglou A, Loupassaki S, Roussis V, Makris AM, Kampranis SC. Production of the forskolin precursor 11 β-hydroxy-manoyl oxide in yeast using surrogate enzymatic activities. Microbial Cell Factories. 2016;**15**:46

[69] Wattanachaisaereekul S, Lantz AE, Nielsen ML, Andresson OS, Nielsen J. Optimization of heterologous production of the polyketide 6-MSA in *Saccharomyces cerevisiae*. Biotechnology and Bioengineering. 2007;**97**:893-900

[70] Mutka SC, Bondi SM, Carney JR, Da Silva NA, Kealey JT. Metabolic pathway engineering for complex polyketide biosynthesis in *Saccharomyces cerevisiae*. FEMS Yeast Research. 2006;**6**:40-47

[71] Cardenas J, Da NA. Engineering cofactor and transport mechanisms in *Saccharomyces cerevisiae* for enhanced acetyl-CoA and polyketide biosynthesis. Metabolic Engineering. 2016;**36**:80-89

[72] Pretorius IS. Synthetic genome engineering forging new frontiers for wine yeast. Critical Reviews in Biotechnology. 2017;**37**:112-136

[73] Shin SY, Han NS, Park YC, Kim MD, Seo JH. Production of resveratrol from p-coumaric acid in recombinant *Saccharomyces cerevisiae* expressing 4-coumarate: Coenzyme A ligase and stilbene synthase genes. Enzyme and Microbial Technology. 2011;**48**:48-53

[74] Shin SY, Jung SM, Kim MD, Han NS, Seo JH. Production of resveratrol from tyrosine in metabolically engineered *Saccharomyces cerevisiae*. Enzyme and Microbial Technology. 2012;**51**:211-216

[75] Sydor T, Schaffer S, Boles E. Considerable increase in resveratrol production by recombinant industrial yeast strains with use of rich medium. Applied and Environmental Microbiology. 2010;**76**:3361-3363

[76] Li M, Schneider K, Kristensen M, Borodina I, Nielsen J. Engineering yeast for high-level production of stilbenoid antioxidants. Scientific Reports. 2016;**6**:1-8

[77] Koopman F, Beekwilder J, Crimi B, van Houwelingen A, Hall RD, Bosch D, van Maris AJA, Pronk JT, Daran J-M. *De novo* production of the flavonoid naringenin in engineered *Saccharomyces cerevisiae*. Microbial Cell Factories. 2012;**11**:155

[78] Eichenberger M, Joanna B, Folly C, Fischer D, Martens S, Simón E, Naesby M. Metabolic engineering of *Saccharomyces cerevisiae* for *de novo* production of dihydrochalcones with known antioxidant, antidiabetic, and sweet tasting properties. Metabolic Engineering. 2017; **39**:80-89

[79] Brown S, Clastre M, Courdavault V, Connor SEO. *De novo* production of the plant-derived alkaloid strictosidine in yeast. Proceedings of the National Academy of Sciences. 2015;**112**:3205-3210

[80] Trenchard IJ, Siddiqui MS, Thodey K, Smolke CD. *De novo* production of the key branch point benzylisoquinoline alkaloid reticuline in yeast. Metabolic Engineering. 2015;**31**:74-83

[81] Hawkins KM, Smolke CD. Production of benzylisoquinoline alkaloids in *Saccharomyces cerevisiae*. Nature Chemical Biology. 2008;**4**:564-573

[82] Galanie S, Smolke CD. Optimization of yeast-based production of medicinal protoberberine alkaloids. Microbial Cell Factories. 2015;**14**:144

[83] Fossati E, Ekins A, Narcross L, Zhu Y, Falgueyret J, Beaudoin GAW, Facchini PJ, Martin VJJ. Reconstitution of a 10-gene pathway for synthesis of the plant alkaloid dihydrosanguinarine in *Saccharomyces cerevisiae*. Nature Communications. 2014;**5**:1-11

[84] Li Y, Smolke CD. Engineering biosynthesis of the anticancer alkaloid noscapine in yeast. Nature Communications. 2016;**7**:1-14

[85] Fossati E, Narcross L, Ekins A, Falgueyret J, Vincent J. Synthesis of morphinan alkaloids in *Saccharomyces cerevisiae*. PLoS One. 2015;**10**:1-15

[86] Galanie S, Thodey K, Trenchard IJ, Filsinger Interrante M, Smolke CD. Complete biosynthesis of opioids in yeast. Science. 2015;**349**:1095-1100

Chapter 5

Non-Conventional Yeasts in Fermentation Processes: Potentialities and Limitations

Dorota Kręgiel, Ewelina Pawlikowska and
Hubert Antolak

Additional information is available at the end of the chapter

http://dx.doi.org/10.5772/intechopen.70404

Abstract

Traditionally the term 'yeast' means *Saccharomyces cerevisiae* and its close relatives. This yeast is used in traditional fermentation processes, mainly for ethanol formation, baking, winemaking and beer production. The classical carbon substrates for typical yeast processes are glucose or sucrose, however, the successful expansion of industrial biotechnology drives research toward the utilization of alternative carbon sources. New technologies require very specific challenges and differ from those found in conventional fermentation processes. Most microbial habitats, especially in modern biotechnological processes, do not provide culture media rich in mono- and disaccharides. They include fermentation environments with various compositions of carbon and energy sources as well as the presence of various cytotoxic compounds which inhibit the growth of industrial yeasts. About 1500 various yeast species have been identified nowadays. Microbiologists and biotechnologists have named all non-*S. cerevisiae* yeasts as 'non-conventional' yeasts. Their features present a potential that can be used for non-conventional processes. Non-*Saccharomyces* strains provide alternative metabolic routes for substrate utilization and product formation. The diversity of these yeasts includes many species possessing useful, and sometimes uncommon, metabolic features potentially interesting for biotechnology. The selected strains of non-conventional yeasts could be used as pure or mixed cultures for improving industrial fermentations.

Keywords: non-*Saccharomyces*, yeasts, fermentation, stress resistance

1. Introduction

Yeasts belong to the most studied microorganisms. More than 1500 species of yeast have been described so far [1]. Many of them have been used in various fermentation processes [2].

Taxonomic analysis of the microflora in active spontaneous fermentations revealed variety of yeasts, but still the predominant genera is *Saccharomyces* [3, 4]. This yeast has become the model organism for research studies and valuable results for numerous eukaryotic cells have been obtained [5]. *S. cerevisiae* was also the first species whose genome was sequenced [6]. The ability of *S. cerevisiae* to conduct metabolic processes under both aerobic and anaerobic conditions, and ethanol production meant that this species has been used for many years as starter cultures for production of bread and numerous fermented beverages [7]. This yeast has also been used in the biofuel industry and for the production of heterologous proteins, human insulin, hepatitis and human papillomavirus vaccines [8].

However, new technological processes, for example, production of second-generation bioethanol, are different from those encountered in conventional fermentation processes. These new technologies pose special challenges. They include fermentations in various environments, with wide spectrum of carbon and energy sources, as well as with significant content of numerous cytotoxic compounds that may inhibit the growth of industrial microorganisms [9, 10]. Strong pressure to improve the economic viability of bioethanol production from waste plant materials makes strains of *S. cerevisiae* rather ineffective in fermentation processes with lignocellulosic hydrolysates. This fact stimulates research to use other non-*Saccharomyces* strains that exhibit broad spectrum of assimilated carbon compounds and higher resistance to cytotoxic compounds.

2. Ethanol production

Ethanol production on the industrial scale has been carried out in the conventional manner using mesophilic strains of *Saccharomyces* spp. The commonly used carbon sources are molasses, beet juice, beet sugar, corn or potato starch. However, these raw materials are expensive and their availability is usually dependent on seasonal productivity. Additionally, the use of plant food such as corn and potatoes in biofuel production is morally and socially controversial. Therefore, diversified actions have been taken to convert a variety of agricultural and forestry wastes, rich in lignocellulosic sugars, into biofuels (**Table 1**).

Biomass	Ethanol yield (litres per dry metric ton)
Hardwood	350
Softwood	420
Corn stover	275–300
Wheat straw	250–300
Sugarcane bagasse	314
Municipal solid waste	170–486

Table 1. Ethanol yields from selected waste biomass [11].

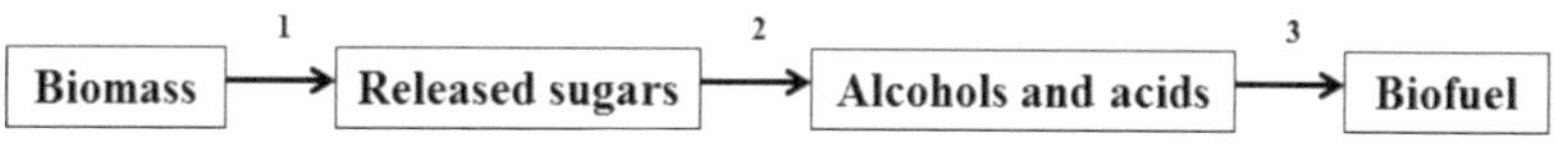

Figure 1. Flow diagram of biofuel production from plant biomass. (1) Pretreatment, (2) fermentation and (3) separation and purification.

According to the Directives 2008/98/EC and Regulation (EU) No 1357/2014, by-products generated can be used directly, without further processing, but wastes may be subjected to recovery, disposed of or liquidated. Proper management of waste biomass is an important issue for environmental protection. However, the use of production residues not only minimizes the negative impact on the environment, but it is also possible to get additional economic benefits [12]. Organic waste from the agroindustry and forestry, according to their physicochemical properties, can be used for the production of bioethanol, butanol, acetone and new chemical building blocks for advanced materials [13, 14] (**Figure 1**).

3. Starch and lignocellulosic biomass

Starch is the carbohydrate accumulated in plants, made up of long chains of glucose units joined by α-1,4 linkages and joined at branch points by α-1,6 bonds. Many microorganisms, including *S. cerevisiae*, are not able to degrade starch since they do not produce starch decomposing enzymes such as α-amylase, β-amylase, pullulanase, isoamylase and glucoamylase. To simplify the fermentation process by eliminating the separate saccharification step, numerous genetically engineered *S. cerevisiae* strains capable of secreting glucoamylase or α-amylase were constructed. However, starch decomposition abilities presented by these yeast strains are usually unsatisfactory because of the limited amounts of secreted amylolytic enzymes [15].

Lignocellulose is the most abundant renewable biomass on earth. It is composed mainly of cellulose, hemicellulose and lignin. Both the cellulose and hemicellulose fractions are polymers of sugars and thereby a potential source of fermentable carbon sources. Hence the interest in research on procedures for chemical degradation of the lignocellulosic structure and for maximization of its decomposition into glucose, xylose and phenolic compounds. The resulting carbon substances can then be assimilated by yeast, which considerably increases the efficiency of biodegradation [16, 17].

Different pretreatment technologies published in public literature are described in terms of the involved mechanisms, advantages versus disadvantages, and economic calculation. Pretreatment technologies for lignocellulosic biomass include biological, mechanical or chemical methods, and their various combinations in particular. It is not possible to define the best pretreatment method because it depends on the type of lignocellulosic biomass and desired products. The acidic (H_2SO_4) or alkali (NaOH) hydrolysis, oxidation techniques (H_2O_2), heat and enzymatic (cellulases, cellobiase and xylanase) treatments are the most frequently used for this purpose [18, 19]. However, each of these methods leads to release of various decomposition products. When fermentable sugars are produced, special attention must be paid to the formation of fermentation inhibitors. Especially the formation of phenolic compounds from lignin degradation, as well as the formation of furfural and 5-(hydroxymethyl)-2-furfural from sugar degradation should be limited by keeping the process parameters: temperature and time as low and as short as possible. Therefore, the choice of the appropriate pretreatment method of plant biomass hydrolysis is the crucial step for effectiveness of fermentation processes [20].

4. *S. cerevisiae* or non-conventional yeasts?

All yeasts are capable of assimilating glucose, almost all such as fructose and mannose, while galactose can also be assimilated by many species. Among the disaccharides, sucrose is the most commonly used. However, in ethanol production, second generation classical yeast *Saccharomyces* spp. are not useful because they are not able to ferment pentoses, exhibit low tolerance to alcohols, acids and solvents. Additionally, they are characterized by high sensitivity to pH changes and cytotoxic compounds: furfural, 5-(hydroxymethyl)-2-furfural and other organic compounds produced during hydrolysis. Limitations of *S. cerevisiae* make the course of new industrial fermentation processes very difficult.

The fuel ethanol production from lignocellulosic materials requires co-fermentation of both hexoses and pentoses, mainly D-xylose and L-arabinose. *S. cerevisiae* cannot utilize pentoses because of the lack of specific metabolic pathways and transport systems. Genomic resources from a variety of microorganisms as well as biological systems combined with mutagenesis have been used to engineer yeast with pentose fermentation abilities [21]. By expressing heterologous D-xylose or L-arabinose pathways, *S. cerevisiae* could obtain the metabolic capacity but this efficiency still needs to be improved [22, 23].

The main strategies for constructing D-xylose-utilizing *S. cerevisiae* include two paths. The first one is XR-XDH pathway, containing D-xylose reductase (XR) and xylitol dehydrogenase (XDH), and converts D-xylose to xylulose. Due to the cofactor imbalance in this pathway, the accumulation of byproduct is the main problem, which needs to be solved. Another one is XI pathway, which only needs to introduce one D-xylose isomerase (XI) that directly converts D-xylose to xylulose. However, the activity of XI still needs to be increased. The xylulose from both pathways could be phosphorylated to xylulose-5-P by endogenous xylulokinase. Subsequently xylulose-5-P can be further entered into the endogenous pentose phosphate pathway (PPP) to produce ethanol [23].

There are also two main L-arabinose metabolic pathways which are both candidates for constructing L-arabinose-metabolic yeasts. L-Arabinose could be converted to D-xylulose-5-phosphate that then enters into PPP. This pathway needs five important enzymes, including aldose reductase, L-arabinitol-4-dehydrogenase, L-xylulose reductase, D-xylulose reductase and xylulokinase. In addition, this pathway contained two reduction reactions which utilize NADPH, two oxidation reactions which generate NADH, and a kinase reaction [23].

Therefore, the construction of stable *S. cerevisiae* strains able to ferment xylose and/or arabinose is not easy. The co-utilization of D-xylose and L-arabinose was obtained in engineered *S. cerevisiae* strain with a high ethanol yield 0.43 g/g of total sugar [24]. Also selected strains of other yeast belonging to *Pichia stipitis* were shown to ferment hydrolysates with ethanol yields of 0.45 g/g of sugar, so commercialization seems feasible for some applications [25].

An additional problem for the simultaneous consumption of pentoses and hexoses is the inhibition of pentose uptake by D-glucose. Researchers have engineered xylose metabolism in *S. cerevisiae* by over-expressing genes for aldose (xylose) reductase, xylitol dehydrogenase and moderate levels of xylulokinase-enabled xylose assimilation and fermentation. The results obtained by Subtil and Boles suggested that co-fermentation of pentoses in the presence of D-glucose can significantly be improved by the overexpression of pentose transporters, especially if they are not inhibited by D-glucose [26]. However, a balanced proportion of NAD(P) and NAD(P)H must be maintained to avoid xylitol production. It was noted that respiration is critical for growth on xylose by both native and recombinant xylose-fermenting yeasts. Reducing the respiration capacity of xylose-metabolizing yeasts increases ethanol production. In studies conducted by Jeffries and Jin, *S. cerevisiae* was engineered for D-xylose utilization through the heterologous expression of genes for aldose reductase, xylitol dehydrogenase and D-xylulokinase and produced only limited amounts of ethanol in xylose medium. It was observed that levels for glycolytic, fermentative and pentose phosphate enzymes did not influence significantly on glucose or xylose under aeration or oxygen limitation. However, expression of genes encoding the tricarboxylic acid cycle and respiration enzymes increased significantly when cells were cultivated on xylose, and the genes for respiration were even more elevated under oxygen limitation. These results suggest that recombinant *S. cerevisiae* does not recognize xylose as a fermentable carbon source. However, the petite respiration-deficient engineered strain produced more ethanol and accumulated less xylitol from xylose [25, 27].

The results obtained by Wang et al. for co-utilization of D-glucose, D-xylose and L-arabinose in engineered *S. cerevisiae* showed that the pentose metabolic capacity is prominently lower than that of D-glucose due to D-glucose-inhibition effect. To alleviate the phenomenon, the pentose metabolic flux can be improved and a pentose specific transporter without inhibition by D-glucose might also be needed [23].

The progress in fermentation of pentose sugars has gone on slow pace as there are few microorganisms known, which are capable of pentose metabolism. While numerous metabolic engineering strategies have been developed in laboratory yeast strains, only a few approaches have been realized in industrial strains. Ethanol yields of more than 0.4 g of ethanol/g of sugar have been achieved with several xylose-fermenting industrial strains with the heterologous xylose utilization pathway consisting of xylose reductase and xylitol dehydrogenase, which demonstrates the potential of pentose fermentation in lignocellulosic ethanol production [28]. In the future, desired perspective is to find organisms that would be able to ferment high density hydrolysates without purification. The genetic and metabolic engineering routes also should be continued. Also a direct or a sequential fermentation system using mixed populations of yeasts needs to be worked out [29].

The interest of microbiologists has been also directed to the use of yeasts belonging to other genera than genus *Saccharomyces* or *Schizosaccharomyces*, commonly called 'non-conventional' yeasts. Due to the information collected on abilities of some of these yeasts, as well as their applications in many fields, their 'unconventional' status may change in the future. Some of the 'non-conventional' yeasts of today will be the 'conventional' yeasts of tomorrow [30]. The similar thesis was given by Sibirny and Scheffers [31]. They highlighted that, since an increasing number of non-conventional yeasts and increasing importance in both fundamental and applied sciences, the term 'non-conventional' is gradually losing significance and usefulness.

There is an enormous biodiversity of non-conventional yeasts. Currently 1500 species have been described although this is only thought to be 1% of yeast that may exist on Earth. These yeasts are phylogenetically diverse and thus may probably harbor industrially relevant traits to augment the currently used *S. cerevisiae*. In addition, due to the carbon substrates utilization range, as well as a poor stress tolerance drawback, there is need to search for novel traits in other yeasts. Therefore, biodiversity is an alternative approach to genetically improved yeasts [32]. Due to the progress in identification and characteristic of a new species found in nature, it is possible to increase the diversity and number of yeasts used in industrial purposes. It is indisputable that the exploration for new species will lead to additional novel technologies, including fermentation of pentoses to ethanol.

A lot of genera different than *Saccharomyces* may also be interesting for their use in specific technological applications. In fact, some species have already attracted researchers in the last years on different aspects: *Kluyveromyces lactis* as a possible utilizer of the residual whey in dairy industries; some methylotrophic yeasts for the production of heterologous proteins; *Yarrowia lipolytica* for its ability to grow on particular substrates and its high protein excretion capacity. As it was mentioned above, transport of carbohydrates into cells is the very important step in yeast metabolism, except in those cases in which di- or trisaccharides are hydrolyzed outside the cell. Transport of monosaccharides such as glucose, fructose or mannose in *S. cerevisiae* is a facilitated diffusion process; however, the situation may be different in other yeasts. For example, in *K. lactis* glucose transport appears to proceed by facilitated diffusion. In *Candida utilis*, the popular 'fodder yeast', glucose appears to be transported by a proton symport when the organism is grown at low glucose concentration [33].

The non-conventional yeasts may overcome many problems related with narrow spectrum of carbon sources assimilation presented by conventional *S. cerevisiae* [15]. Some non-conventional yeasts show many uncommon, metabolic features potentially interesting to biotechnology. Non-conventional yeasts represent the vast majority of genera and species so far described. Several yeast species are diverged by evolution from *S. cerevisiae* and possess several unique genes and growth characteristics to withstand different stress conditions [34]. These exceptional strains are able to utilize various sources of carbon such as starch, cellulose, raffinose, arabinose, xylose and sugar alcohols (xylitol, sorbitol, mannitol, etc.) [8, 35].

At least 22 yeast strains have been shown to produce some ethanol from D-xylose. However, only six strains such as *Brettanomyces naardenensis, C. shehatae, C. tenuis, Pachysolen tannophilus, P. segobiensis* and *P. stipitis* are able to produce significant amounts of ethanol, and of these, only three: *C. shehatae, P. tannophilus* and *P. stipitis* have been studied extensively [36, 37].

The production systems exploiting some non-*Saccharomyces* yeasts have one important advantage—they are not pathogenic organisms received the 'generally recognized as safe' (GRAS) designation from the Food and Drug Administration (FDA) [38–40].

The non-conventional yeast systems may have several beneficial traits like ethanol tolerance, thermotolerance, inhibitor tolerance, genetic diversity, etc. However, not all non-conventional yeasts possess these important characteristics. Currently, studies on non-conventional yeasts concern limited number of species like *Hansenula polymorpha, K. lactis, P. pastoris* and *Y. lipolytica* [21, 40]. However, more non-conventional yeasts are worth the special attention. For example, amylases from non-conventional yeasts were found to have the ability to hydrolyze starch. These interesting enzymes are α-amylase and glucoamylase from *Debaryomyces (Schwanniomyces) castelli*, glucoamylase from *Saccharomycopsis fibuligera* and *C. antarctica*, α-amylase from *Cryptococcus* sp., *C. lusitaniae, C. famata*, amylopullulanase from *Clavispora lusitaniae* and pullulanase from *Aureobasidium pullulans*. There is a number of reviews on bacterial and fungal amylases and their applications. They clearly indicate that α-amylases and pullulanase from yeasts are one of the most popular and important forms of industrial amylases. Non-conventional yeasts were studied as both free and immobilized cells for production of amylolytic enzymes [41–44] (**Photo 1**).

Fermentation trials with immobilized conventional *S. cerevisiae* and non-conventional cells *D. occidentalis* showed that both tested yeasts are able to adapt to the specific conditions inside carrier materials. Nevertheless, the mechanical endurance of alginate carriers, commonly used in yeast immobilization, shows better applications in industrial fermentation especially with non-conventional yeasts. In the case of fermentative yeast, *S. cerevisiae* alginate beads may be destroyed, as a result of intense CO_2 formation [44]. Furthermore, *Debaryomyces* spp. ability to tolerate and decompose both phenols and polyphenols at concentrations that are highly toxic to bacteria and other yeast species, demonstrated that these yeasts may be an attractive system for biofuel production from renewable starch sources [15]. It is worth to note that methylotrophic yeasts belonging to *Hansenula, Candida, Pichia* and *Torulopsis* genera are able to metabolize monocarbonic compounds like methanol and formaldehyde [45]. As

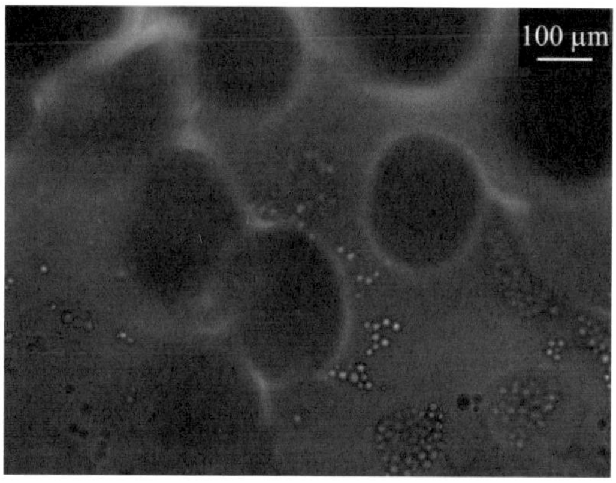

Photo 1. Amylolytic non-conventional yeast *D. occidentalis* immobilized in alginate (author: Dorota Kregiel).

a result, the use of non-conventional yeasts allows for the utilization of various waste plant biomass, and—which is worth emphasizing—to receive the post-fermentation yeast biomass rich in protein and amino acids [39, 46, 47].

Non-conventional yeasts are large but not fully known, diverse group of microorganisms. Yeast species other than *Saccharomyces* spp., in addition to the previously mentioned ability to use the complex substrates, exhibit features particularly important in the industrial processes—thermotolerance and tolerance to the presence of chemical inhibitors. The majority of non-conventional yeasts have been isolated and characterized as microflora of spoiled food and beverages [48, 49]. It can be assumed that some non-conventional yeasts have developed specific mechanisms to survive in various natural environmental conditions. Therefore, it is believed that most species of non-conventional yeasts have acquired specific mechanisms that are not included in the classical model yeast *S. cerevisiae* [49–51]. It is worth to explore the molecular basis of their tolerance to numerous environmental stress such as increased osmotic pressure, high concentrations of ethanol, high temperature or the presence of toxic compounds.

The eukaryotic microorganism most studied for its tolerance is still *S. cerevisiae*. However, this yeast is rather sensitive and not able to adapt to 'non-regular' conditions. For example, some species like *D. hansenii* or *Hortaea werneckii* have been isolated from natural hypersaline environments. Therefore, these non-conventional yeasts are more suitable model organisms to study halotolerance in eukaryotes than *S. cerevisiae* [52]. It should be also noted that non-conventional yeasts show a high growth rate in fermentation processes, and are capable of producing important enzymes. For example, *D. occidentalis* is capable of secretion of α-amylase and glucoamylase, and *K. marxianus* of producing intracellular lactase, intra and extracellular pectinase and intra and extracellular inulinase [42, 44, 53–55]. These examples show the important potential of non-conventional yeasts that can be utilized in use of various waste materials.

5. Osmotolerance

Yeast cells are exposed to osmotic stress during industrial fermentations. The processes carried out in media with significant levels of saccharides above 300 g/L, need osmotolerant yeasts in particular [56–58]. Accordingly, there is a growing interest among microbiologists and technicians to obtain yeast strains that are able to grow in environments with high concentrations of salts or saccharides. The molecular mechanisms for responsibility of osmotolerant *S. cerevisiae* strains have been described extensively in available literature [59, 60]. *S. cerevisiae* remains the model organism to study the molecular basis of important physiological features, but researchers have isolated and identified non-conventional osmotolerant yeasts belonging to *Zygosaccharomyces rouxii* [48, 61].

Z. rouxii is known for its high tolerance to osmotic stress, which is thought to be caused by sets of specific genes. Important differences were found for salt tolerance and assimilation of glycerol in comparison to *S. cerevisiae*. *Zygosaccharomyces* strains show a higher resistance to salts, higher glycerol production and are able to assimilate glycerol. Under conditions of osmotic stress, the glycerol production in *Z. rouxii* strains may be much lower than in *S. cerevisiae*, which suggests the presence of a system that efficiently retains glycerol inside *Z. rouxii* cells [48, 61, 62].

D. hansenii is also one of the most halotolerant species. This yeast was isolated from saline environments sea water, concentrated brines and salty food. It can grow in media containing as high as 4 M NaCl, while the growth of *S. cerevisiae* is limited in media with more than 1.7 M NaCl [63].

The adaptation of yeast cells to osmotic stress is a complex mechanism that combines network regulatory genes and signaling pathways that may vary depending on the species and osmotic agent in the surrounding environment [64]. Generally, the behavior of *Z. rouxii* cells resembles the activity of *S. cerevisiae* in the transport of Na^+ ions from yeast cell, while halotolerant yeast *D. hansenii* accumulates sodium ions inside its cells.

The results obtained by Gonzalez-Hernandez et al. confirmed that *D. hansenii* grows better in the presence of moderate concentrations (0.6 M) of NaCl and KCl than in the absence or at higher salt concentration. Therefore *D. hansenii* can be considered moderate halophile yeast [65]. This ability is associated with the accumulation of high concentrations of K^+ or Na^+. For this reason *D. hansenii* has been called a 'sodium-includer' [66, 67]. The mechanism of the adaptation is probably an intrinsic resistance to the toxic effects of cations, not observed in other yeasts, particularly *S. cerevisiae* [65]. The problems, how yeasts regulate the intracellular ion concentration, and how ions are tolerated by enzymes promoting survival, remains controversial [67, 68]. In *D. hansenii*, the vacuolar concentration of Na^+ was described to be equal to the one of cytoplasm, while in *S. cerevisiae* the differences between these concentrations were described [67].

Z. rouxii integrates general and osmoticum-specific adaptive responses under sugar and salts stresses, including regulation of Na^+ and K^+-fluxes across the plasma membrane, modulation of cell wall properties, compatible osmolyte production and accumulation and stress signaling pathways [69, 70]. According to Leandro et al., *Z. rouxii* is capable of growing in osmolarity of 3 M NaCl and glucose concentrations of 90%, due to the presence of unique transporters in plasma membrane which is higher than *S. cerevisiae* [62]. Dakal et al. described internal reactions that occur in yeast cells under different osmotic agents. They suggested that sugars and polyols modify the osmotic pressure, while salts induce changes in both osmotic pressure and ionic homeostasis [70].

According to Pribylova et al., the less osmotolerant yeasts strain possesses a more rigid cell wall than the more osmotolerant ones. They suggested that the differences in the osmotolerance are related to resistance to the lysing enzymes—lyticase and zymolyase, cell-wall polymer content and cell wall micromorphology [69].

Availability of genome sequence of osmotolerant and halotolerant strains may open up new perspectives in this direction [71].

6. Thermotolerance

Thermotolerance of yeast cells is a highly desirable feature for fermentation processes. Efficient process for bioethanol production from lignocellulosic substrates requires relatively high temperatures

(~50°C) for conducting the enzymatic hydrolysis of biomass before fermentation [72]. Moreover, fermentations carried out at high temperatures significantly reduce the costs of cooling, as well as the risk of microbial contamination [73]. A limited temperature tolerance in yeast *S. cerevisiae*, with the optimal range of 25–37°C, increases the overall cost of ethanol production [74, 75]. Therefore, in order to achieve efficient fermentation at high temperatures, thermotolerant microorganisms may be used. These strains are not only able to survive, but also to produce ethanol efficiently [75, 76]. Non-conventional strains of *K. marxianus* show ability to ferment carbon sources at the temperature of 45°C. Thermotolerance, a broad enzymatic activity and fermentation ability in high concentration of saccharides makes the yeast *K. marxianus* a good material to conduct various fermentation processes [77]. Also other non-conventional yeast species-like *Ogataea polymorpha* (syn. *H. polymorpha*) have been found to ferment xylose at 45°C [78].

Yeast thermotolerance is the result of many factors, including trehalose, heat shock proteins, ATPase, the ubiquitin-proteasome pathway, gene expression responses and heat-induced antioxidant defenses [79]. Some processes may be specific to basal thermotolerance, others may be induced during acquired thermotolerance, and many may be involved in both. High temperatures are known to affect membrane-linked processes due to alterations in membrane fluidity and permeability. Enzymes are also sensitive to higher temperatures. Heat-induced protein denaturation can lead to imbalance in metabolic pathways or to complete enzyme inactivation. These changes lead to the production of active oxygen species and, consequently, heat-induced oxidative stress [80].

The best-characterized aspect of acquired thermotolerance is the production of heat shock proteins (HSPs) consisting of a helix-turn-helix class DNA binding domain, a leucine zipper domain required for trimerization, and a carboxy-terminal transcriptional activation domain. In *S. cerevisiae*, heat shock factor (HSF) is encoded by a single, essential gene, *HSF1*. It was documented that Hsf1p protein from *S. cerevisiae* and HSF from yeast *K. lactis* both contain a unique transcriptional activation domain amino-terminal to the DNA binding domain. Hsf1p appears to be primarily responsible for production of protein chaperones during heat shock [81]. At higher temperatures, organisms induce massive transcription and translation of HSPs. These proteins are proposed to act as molecular chaperones to protect cellular proteins against irreversible heat-induced denaturation and to facilitate refolding of heat-damaged proteins. Genetic evidence established that the Hsp100 family proteins are essential for the acquisition of thermotolerance [82].

The major role for the pathway in heat shock response is mediated by expression of genes required for the synthesis and degradation of the disaccharide trehalose. Originally thought to function as a storage carbohydrate, trehalose accumulates to extremely high levels in stationary phase cells. Logarithmic-phase cells have very low levels of trehalose, which are rapidly increased upon stress exposure. This acts as cytoprotectant, blocking thermally induced protein aggregation. Importantly, trehalose-stabilized proteins are maintained in a partially folded state, ready for reactivation by protein chaperones. Accordingly, the continued presence of trehalose inhibits protein refolding. Stress recovery therefore requires reduction of cellular trehalose levels. Trehalose can thus be considered a chemical chaperone for protein folding with properties remarkably similar to the chaperone Hsp104p – the ability to stabilize unfolded proteins and prevent aggregation [83].

It was documented that temperature affects both growth and ethanol tolerance. Decreasing temperature decreases membrane fluidity; increasing temperature increases membrane fluidity. Yeasts are able to adapt to low temperatures by increasing the proportion of cis-unsaturated fatty-acyl groups in lipids forming cell membranes. Physical principles suggest that fluidity would decrease as the ratio of saturated to unsaturated fatty acids increases because desaturation introduces a bend in the fatty acid chain. However, the majority of fatty acids in the membranes of *S. cerevisiae* are unsaturated, so other factors may be more important. It was found that the unsaturation level of *S. cerevisiae* cellular fatty acids increases at both sublethal or supraoptimal temperatures. On the other side, it was noted that the high content of unsaturated fatty acids is rather result from activation of oxygen-consuming desaturase activity. Membrane fluidity is also affected by the ratios of cell lipids and proteins. These vary with the yeast strain and the conditions under which it is cultivated [37, 84, 85].

Ethanol also affects membrane fluidity, but through different mechanisms. The presence of alcohols, results in the decrease of the temperature required for maximal activation of heat-shock genes, and the concentration of alcohol needed decreases with alcohol chain length. Ethanol is thought to alter membrane organization and permeability by entering the hydrophobic interior and increasing the polarity of this region [37].

The plasma membrane proton pump (H^+-ATPase) of yeast couples ATP hydrolysis to proton extrusion, thereby providing the means for solute uptake by secondary transporters and for regulating cytoplasmic pH. By pumping protons out of the cytoplasm, the H^+-ATPase acidifies the external medium, and makes the cytoplasm relatively alkaline. *S. cerevisiae* possesses two isoforms of this enzyme Pma1 and Pma2. They are 89% identical at the protein level, but they exhibit different activation, kinetic and regulatory properties, which may suggest their different functions. The specific activity of Pma1 increases with growth temperature. However, the increase in activity following stress is not attributable to synthesis of new protein, but rather to activation of the existing enzyme. Additionally, in *S. cerevisiae*, protein Hsp30 is a stress-inducible regulator of ATPase activity. Hsp30 is induced by heat shock, ethanol exposure, severe osmostress, weak organic acid exposure and glucose limitation. Hsp30 induction downregulates stimulation of H^+-ATPase caused by stress. There were also extensive studies of ATPase activity in non-conventional yeast *P. stipitis*. The enzyme from this yeast attained its highest activity at 35°C. It is unclear whether ATPase activity in *P. stipitis* involves one protein or two, as in the case of *S. cerevisiae*. Plasma membrane ATPase activity is essential for basal heat resistance. Moreover, thermotolerance is enhanced by prior exposure to stress. Pre-stressed cells are able to protect the proton gradient longer than cells that have not adapted to heat [86].

High-temperature stress causes multiple changes in the cell that ultimately affect protein structures and function, leading to inhibition of cell growth or cell death. The denatured or aggregated proteins in live cells may be degraded via the ubiquitin proteasome pathway (UPP). It is the one of main defense strategies to ensure survival in stress conditions [87]. This is ATP-dependent process, and timely destruction is vital for controlled cell division, as well as proteins unable to fold properly within the endoplasmic reticulum. The UPP is carried out by three classes of enzymes. A 'ubiquitin activating enzyme' (E1) forms of a thio-ester bond with ubiquitin that is a highly conserved 76-amino acid protein. The next reaction allows binding of ubiquitin to a 'ubiquitin conjugating enzyme' (E2), followed by the formation of the isopeptide bond between C-terminus of ubiquitin and the lysine rest by 'ubiquitin ligase' (E3) action. The UPP selectively eliminates abnormally folded or damaged proteins that have arisen by missense or nonsense mutations, biosynthetic errors, or damage by oxygen radicals or by denaturation, especially at high temperatures [88].

The mechanisms of yeast thermotolerance are largely controlled through the activation and regulation of specific stress-related genes involved in the synthesis of specific compounds that protect the organism from high-temperature stress. Elucidation of the function of these genes and/or proteins will give insight into the various mechanisms underlying yeast response to high-temperature stress, providing useful information to improve bioethanol production at higher temperatures.

Genetic data indicate that different genes contribute to heat tolerance at different stages of the plant life cycle and that different genes may be essential for basal and acquired thermotolerance [82]. Studies conducted by Gibney et al. have shown that gene deletions may also lead to higher thermosensitivity. Functional analysis of some identified genes confirmed that metabolism, cellular signaling and chromatin regulation play key role in controlling of yeast thermotolerance. However, the molecular mechanism of these actions remains still imprecise. They suggest that survival after heat shock depends on a small number of genes that function in assessing the metabolic health of the cell and/or regulate its growth in a changing environment [89]. To understand the mechanism of thermoadaptation, Shui et al. performed proteomic analysis for both parental and evolved strains of *S. cerevisiae*. They showed that some proteins were differentially regulated at heat-stress conditions in the parental and evolved strains. Additionally, the proteomic response of the industrial strains adapted to stress conditions was substantially different in comparison to the response of laboratory yeast to unexpected heat stress [90].

7. Fermentation activity and ethanol tolerance

Oxygen is one of key factors in regulation of fermentation in yeast. According to the role of oxygen in their metabolism, yeasts can be classified as: (a) obligatory aerobic, with only respiratory metabolism; (b) facultative fermentative or respiro-fermentative, displaying both respiratory and fermentative metabolism and (c) obligatory fermentative [91]. Although the majority of yeast species described so far is able to ferment sugars into ethanol and carbon dioxide, most of the respire-fermentative yeasts do not grow well under strictly anaerobic conditions [92].

Van Dijken and Scheffers explained the central role of two redox couples NAD^+/NADH and $NADP^+$/NADPH in the metabolism of sugars by yeasts. NADH is preferentially used in dissimilatory metabolism, whereas NADPH is generally required for assimilatory reactions. In *S. cerevisiae*, *C. utilis* and probably in the yeasts in general, NADH and NADPH cannot be interconverted owing to the absence of a transhydrogenase activity [93].

Barnett [94] described 678 yeast species, and around 60% are considered to be fermentative on the basis of taxonomic tests such as gas production (Durham tubes) in laboratory conditions. However, this number is even higher since, under certain conditions, some of those species considered as non-fermentative are also able to ferment glucose. The ability to ferment glucose under oxygen limitation turns out to be a common feature of the different yeast species, but the capability of growth under anaerobic conditions is not widespread among these microorganisms. In fact, only very few yeast species are capable of fast growth under those conditions and *S. cerevisiae* stands out as the yeast generally acknowledged as a facultative anaerobe. Anaerobic growth is associated with a low energy yield compared with that observed under complete oxidative processes [94].

The Pasteur and the Crabtree effects are the examples of special competition between respiration and fermentation of glucose [93]. The Pasteur effect refers to an activation of anaerobic glycolysis in order to meet cellular ATP demands owing to the lower efficiency of ATP production by fermentation compared with respiration. The Crabtree effect is currently defined as the occurrence of alcoholic fermentation under aerobic conditions. These two regulatory effects are very important in industrial fermentation [95, 96].

S. cerevisiae utilizes glucose by fermentative pathway (Crabtree positive) and some non-conventional yeasts like *K. lactis*, *P. pastoris* and *Y. lipolytica* are predominantly oxidative (Crabtree negative). However, among non-conventional yeasts are also Crabtree-positive ones. *S. cerevisiae* shows tolerance and good adaptation to high concentrations of ethanol. It was found that *S. cerevisiae* cells grown in the presence of ethanol appear to increase the amount of monounsaturated fatty acids in cellular lipids [97]. However, several non-conventional yeasts such as *Dekkera bruxellensis*, *P. kudriavzevii*, *Torulaspora delbrueckii* or *Wickerhamomyces anomalus* show quite good fermentation abilities and similar levels of ethanol tolerance in comparison to *S. cerevisiae* [98–103]. Especially Crabtree-positive *D. bruxellensis* strains are able to remain viable in fermentation media containing up to 16% ethanol. It has been shown that the yield of ethanol formation by *D. bruxellensis* in batch culture under anaerobic conditions is comparable with conventional yeasts. Additionally, *D. bruxellensis* shows the ability to 'compete' with conventional yeasts in industrial conditions, presumably due to the predominance of *S. cerevisiae* in the assimilation of nitrates [101, 102].

Several attempts were initiated to increase ability of yeast fermentation or to convert Crabtree-negative yeasts into Crabtree-positive for improving ethanol fermentation efficiency. Schifferdecker et al. created a metabolically engineered strain *D. bruxellensis* by increasing its fermentation capability. The gene encoding for alcohol dehydrogenase was overexpressed under the control of highly active *TEF1* promoter. As result, the improved strain produced 1.4–1.7 times more ethanol than the parental yeast [104]. Other unconventional strain of *K. lactis* was constructed as a mutant in the single gene encoding for a mitochondrial alternative internal dehydrogenase. This strain showed unaffected rate of exogenous NADH oxidation, but this mutation shifted the metabolism from respiration to fermentation. As a consequence, the mutant of *K. lactis* showed the increased rate of ethanol production [105].

Cost-effective fermentation depends on, among other factors, rapid and high yielding conversion of carbohydrates to ethanol, which in itself depends on improvements in the survival and performance of yeast cells under industrial conditions. Conventional *S. cerevisiae* is responsible for industrial alcoholic fermentation. On the other hand, most non-conventional yeasts that do not show such regulatory effect, which does not allow for efficient ethanol production in industrial conditions. Therefore, in traditional fermentation processes (beer production and winemaking), the non-*Saccharomyces* yeasts, initially present in fermentation medium at high numbers (ranging from 10^3 to 10^5 cells/ml), grow only during the early stages (up to 4–5% v/v of ethanol) and they are soon overtaken by strongly fermentative *S. cerevisiae* strains that complete the fermentation process [92].

Ethanol is well known as an inhibitor of microbial growth. Large concentrations of ethanol can be toxic to yeasts. Ethanol in low amounts inhibits cell division, decreases cell volume and specific growth rate, while high ethanol concentrations reduce cell vitality and increase cell death [106]. Ethanol also influences cell metabolism and macromolecular biosynthesis. The main results of these changes are production of heat shock-like proteins, low rate of RNA and protein accumulation, numerous petite mutants, denaturation of intracellular proteins and reduction of glycolytic enzymes activity. The response of yeasts to ethanol stress is complex, involving various aspects of cell sensing, signal transduction, transcriptional control, protein-targeting, accumulation of protectants and increased activity of repair functions. The efficiency of these processes in a given yeast strain determines its worth in industrial processes [107].

8. Furan and acetate tolerance

The use of hydrolysates obtained from plant biomass for the production of second generation bioethanol may be very problematic. In the pre-treatment processes, a lot of by-products toxic to the yeast cells may be formed [10]. The composition and concentration of inhibitory compounds is variable and depends on the type of lignocellulosic raw material and the method of its pretreatment [108]. Generally, after the pre-treatment and enzymatic hydrolysis of the hemicellulose fraction, hexoses: D-glucose, D-galactose, D-mannose and D-rhamnose and pentoses: D-xylose and L-arabinose are obtained [109, 110]. However, under high temperature and pressure, hexoses and pentoses may be degraded to 5-(hydroxymethyl)-2-furfural and furfural. The harmful effects of these compounds, even at low concentration, have been confirmed. RNA, DNA, proteins and membranes of yeast cells are particularly sensitive [111, 112].

The removal of toxic compounds from the fermentation medium is usually very expensive. Therefore, in order to improve the fermentation processes, the use of furan-tolerant yeast strains is more practical. Scientists recognize the molecular basis of cell tolerance to furan and its derivatives for model yeast *S. cerevisiae*. It has been found that *SIZ1* gene encoding the ligase E3, can bring the significant increase in tolerance to furfural. Some non-conventional yeast species, namely *W. anomalus, P. kudriavzevii, C. stellata, C. ethanolica, P. fermentans* and *Z. bailii*, show good tolerance to furfural and its derivatives. For example, the resistance of *P. kudriavzevii* to hydroxymethyl reaches up to 7 g/L [103].

The tolerance to weak acids is essential in the second generation bioethanol production. During the pretreatment of the lignocellulosic feedstock, released hemicellulose acetyl groups form acetic acid in the concentration of 5–10 g/L [113, 114]. It is known that weak acids exhibit cytotoxic effects. These compounds are transported through the cell membrane into the yeast cells by passive diffusion in non-dissociated form. In the yeast cells they are subject to dissociation, and protons are accumulated in the cytoplasm, causing acidification of cytosol [115–118]. In this case the cell metabolism slows down significantly by inhibiting glycolytic enzymes and NADH dehydrogenase [119–122]. Low intracellular pH inhibits the growth of yeasts, the adaptive phase increases and consequently, the efficiency of ethanol production decreases [123, 124]. Therefore, the use of yeast strains resistant to weak acids is essential for industrial production of bioethanol. Non-conventional yeast *Z. bailii* has been described as the most resistant to acetic acid. This yeast can grow at the concentration as high as 24 g/L, while conventional *S. cerevisiae* shows sensitivity at 9 g/L of acetate [125].

9. Mixed populations and biocontrol

S. cerevisiae are able to produce high concentrations of ethanol reaching approximately 20% (v/v) but in conventional media. This yeast shows high fermentation rates, whereas they are unexpectedly less tolerant to high concentrations of ethanol and other toxic compounds. That is the reason why several ethanol-tolerant yeasts are used in industrial fermentations.

The profusion of selected starter cultures has allowed the more widespread use of inoculated fermentations, with consequent improvements to the control of the fermentation process, and the use of new biotechnological processes. Mixed fermentations using controlled inoculation of *S. cerevisiae* starter cultures and non-*Saccharomyces* yeasts represent a feasible way toward improving the complexity and enhancing the particular and specific characteristics of fermentation products [126–128].

Mixed cultures with different yeasts also provide an advantage in bioethanol production. In starchy media, using raw unhydrolysed starch in a single-step fermentation, ethanol production by a co-culture of *S. diastaticus* and *S. cerevisiae* was 24.8 g/L. This was 48% higher than the yield obtained with the monoculture of *S. diastaticus* (16.8 g/L). In another coculture fermentation with *Endomycopsis capsularis* and *S. cerevisiae*, maximum ethanol yield was 16.0 g/L, higher than *E. capsularis* the yield with the monoculture [129].

In second-generation ethanol production, xylose and arabinose are the significant fraction of lignocellulosic biomass. Therefore, their utilization is essential for a feasible bioethanol production process. The selection of yeast strains for the fermentation of pentoses has a large effect on ethanol yield [130]. The naturally xylose-fermenting non-conventional yeasts such as *C. shehatae* and *P. stipitis* have been widely studied because of their ability to ferment xylose into ethanol [131]. *P. stipitis* is considered as a promising strain because it can ferment a wide range of sugars, including cellobiose. *Candida* species have been shown to ferment D-xylose to ethanol as the major product. Strain improvement by mutation is one of the best methods to increase the ethanol yield, and in this case, two strains capable of producing significantly higher ethanol yields than the parental strains were obtained [132].

The influence of non-*Saccharomyces* yeasts on fermentation processes was studied and their biotechnological potential was evaluated. The industrial yeast market, which was historically focused on *S. cerevisiae*, now offers *S. cerevisiae*/non-*Saccharomyces* multi-starters. However, the development of these mixed populations requires knowledge about possible interactions between yeast strains. Considering the use of mixed populations, the special attention should be paid not only to the selection of the proper assimilation-competent strains, their inoculation, culture media, but also to the interactions between these yeast monocultures. The interesting results were obtained by Yamaoka et al. [128]. This research was carried out to investigate the influence of non-*Saccharomyces* yeast, *K. lactis*, on metabolite formation and the ethanol tolerance of *S. cerevisiae* in mixed cultures in synthetic minimal medium containing 20% glucose. It was noted that co-cultivation of *K. lactis* seems to prompt *S. cerevisiae* to be ethanol tolerant by forming protective metabolites such as glycerol.

In turn, studies on mixed cultures *S. cerevisiae*/*T. delbrueckii* showed that physical contact between yeast cells induced rapid death of *T. delbrueckii*. This phenomenon was previously described as a cell-cell contact mechanism. However, when these yeast cultures were physically separated from each other, the sensitive strain of *Torulaspora* sp. kept its viability [133].

The mixed yeast populations have been explored not only for improvement of ethanol yield but also as biological control—an alternative to the use of synthetic chemicals for prevention of microbial spoilage. The possibility of using the selected antagonistic yeasts against undesirable spoilage microorganisms is the subject of interest for both scientists and technologists. The presence of undesired microflora may lead to significant reduction in the efficiency of biotechnological processes. Non-conventional yeasts, characterized by antagonistic activity against spoilage microflora include genera *Pichia*, *Candida*, *Aureobasidium*, *Metschnikowia* and *Debaryomyces*. The interactions between microorganisms have been described in numerous scientific studies [126, 127, 134–136]. Industrial yeast strains, due to the high reproductive potential and rich enzymatic equipment, have the ability to colonize fermentation environments rapidly. The presence of microbial contamination not only reduces available nutrients for industrial microorganisms, but also reduces the potential living space. Low nutrient availability is one of the most important mechanisms of competition between yeast strains.

The killer phenomenon was first observed in yeast *S. cerevisiae* in the 1960s of the last century. However, the killer features have also been found in representatives of non-conventional yeasts belonging to genera *Debaryomyces*, *Pichia*, *Kluyveromyces*, *Candida*, *Cryptococcus*, *Ustilago*, *Rhodotorula*, *Williopsis*, *Torulopsis*, *Zygosaccharomyces*, *Hansenula* and *Hanseniaspora*. Killer protein toxins show specific activity spectra dependent on pH level, temperature and aeration conditions. These toxins differ in resistance to proteolytic enzymes, chemicals, pH, and they are mutually antagonistic. The impact of killer yeasts to sensitive yeast cells include killer protein receptors on the cell wall of sensitive cells. The consequences of killer toxin binding to cell wall are physiological changes that lead to death of the sensitive cells. Initially, there is a breakdown of amino acids and proton gradient, leakage of potassium ions from ATP, reduction of metabolite levels and the destruction of the pH gradient. All these processes lead to a gradual death of sensitive yeast cells [4, 137, 138].

It has been found that yeast strains *Metschnikowia pulcherrima* have a great potential to be a leading natural and biological control against a broad spectrum of pathogens [139–141].

http://dx.doi.org/10.5772/intechopen.70404

M. pulcherrima forms pulcherimic acid, which is accumulated in growth medium and forms red pulcherrimin—a chelate complex with Fe(III) ions (**Photo 2**).

It has been shown that the antibacterial and antifungal activity of yeast depends on pulcherrimin formation [139]. Therefore, strains that produce large amounts of pulcherrimin are of great interest to engineers and microbiologists, as biocontrol agents inhibiting growth of pathogenic bacteria, yeasts and molds. This substance may be an alternative to antibiotics and fungicides. Oro et al. evaluated *M. pulcherrima* for the antimicrobial activity against numerous yeast strains belonging to *Pichia, Candida, Hanseniaspora, Kluyveromyces, Saccharomycodes, Torulaspora, Brettanomyces* and *Saccharomyces* genera [141]. *M. pulcherrima* displayed a broad and effective antimicrobial action on undesired wild spoilage yeasts (*Brettanomyces/Dekkera* spp., *Hanseniaspora* spp., *Pichia* spp.). Interestingly, the antimicrobial activity of *M. pulcherrima* did not have any influence on the growth of *S. cerevisiae*. The oxygen availability strongly influences population dynamics in mixed populations of conventional and non-conventional yeasts. Additionally, in the presence of non-*Saccharomyces* yeasts, species-specific chemical volatile profiles were noted, in particular increases in some higher alcohols and medium chain fatty acids. This data show the potential use of selected *M. pulcherrima* strains in controlled multi-starter fermentations with *S. cerevisiae* starter cultures [142, 143].

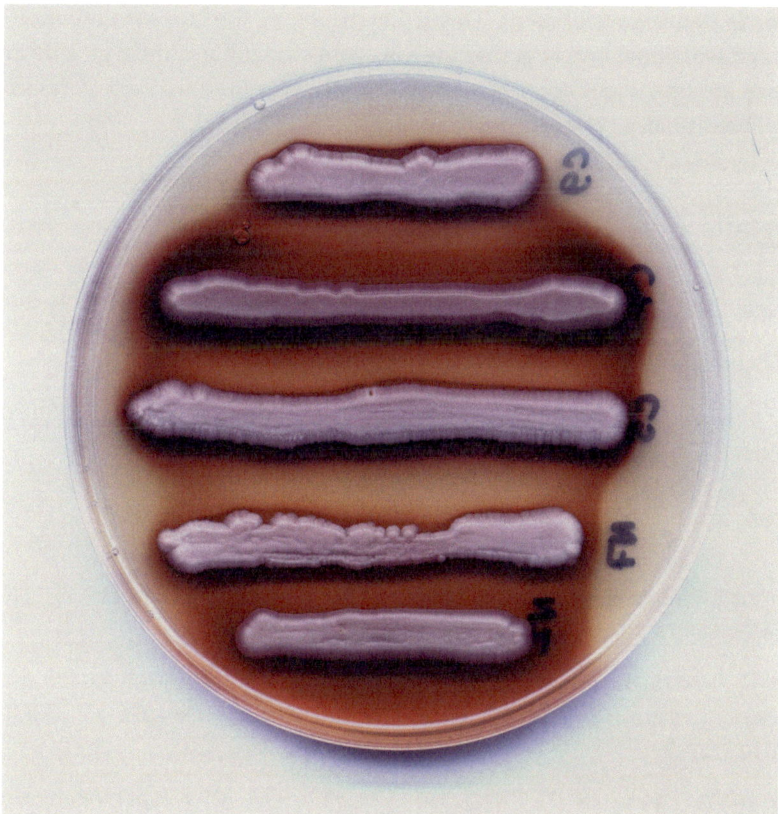

Photo 2. Pulcherrimin formation by *M. pulcherrima* on YPD agar with Fe(III) ions (author: Ewelina Pawlikowska).

10. Conclusion

The conventional yeast *S. cerevisiae* is the best known species used in numerous industrial high-tech processes. However, in new technologies, including second generation ethanol production, the use of this yeast may encounter a number of difficulties. Research studies suggest that it is possible and even necessary to use selected non-conventional yeast strains to increase the use of carbon sources as well as to improve the economic effects of ethanol production from plant waste materials. It is worth paying attention to one more aspect—many species of non-conventional yeasts produce unique biocontrol compounds, which can be seen as an additional valuable feature for conducting fermentation processes. These yeasts may find use as monocultures or mixed complementary populations. Although the exploration of existing natural biodiversity of non-conventional yeasts is attractive, the major bottleneck is that industrially applicable traits are not commonly found in nature. However, there are multiples of classical approaches to develop strains with improved phenotypes such as mutagenesis, sexual hybridization, genetic modification, adaptive evolution and other emerging tools. Among them, non-genetic modification, adaptive evolution, is preferable; as the use of strains developed using genetic methods in the food industry remains controversial. In addition, such a traditional phenotype improvement based on random appearance of adaptive mutations based on selective regimes requires no prior knowledge of the genetic background of the strains is under development. This is important, as the current limitation in applications of non-conventional yeasts is that they are less studied and their genetic architectures and pathways are less understood. Therefore, we can conclude that era of research on non-conventional yeasts has just begun.

Author details

Dorota Kręgiel*, Ewelina Pawlikowska and Hubert Antolak

*Address all correspondence to: dorota.kregiel@p.lodz.pl

Institute of Fermentation Technology and Microbiology, Lodz University of Technology, Poland

References

[1] Boekhout T. Biodiversity – Gut feeling for yeasts. Nature. 2005;**434**:449-451. DOI: 10.1038/434449a

[2] Sicard D, Legras JL. Bread, beer and wine: Yeast domestication in the *Saccharomyces sensu stricto* complex. Comptes Rendus Biologies. 2011;**334**:229-236. DOI: 10.1016/j.crvi.2010.12.016

[3] Bokulich NA, Thorngate JH, Richardson PM, Mills DA. Microbial biogeography of wine grapes is conditioned by cultivar, vintage, and climate. Proceedings of the National Academy of Sciences of the United States of America (PNAS). 2014;**111**:139-148. DOI: 10.1073/pnas.1317377110

[4] Steensels J, Verstrepen KJ. Taming wild yeast: Potential of conventional and nonconventional yeasts in industrial fermentations. Annual Review of Microbiology. 2014;**68**:61-80. DOI: 10.1146/annurev-micro-091213-113025

[5] Bostein D, Chervitz SA, Cherry JM. Yeast as a model organism. Science. 1997;**277**:1259-1260. DOI: 10.1126/science.277.5330.1259

[6] Goffeau A, Barrell BG, Bussey H, Davis RW, Dujon B, Feldmann H, Galibert F, Hoheisel JD, Jacq C, Johnston M, Louis EJ, Mewes HW, Murakami Y, Philippsen P, Tettelin H, Oliver SG. Life with 6000 genes. Science. 1996;**274**:546-567. DOI: 10.1126/science.274.5287.546

[7] Querol A, Fernandez-Espinar MT, Del Olmo M, Barrio E. Adaptive evolution of wine yeast. International Journal of Food Microbiology. 2003;**86**:3-10. DOI: 10.1016/S0168-1605(03)00244-7

[8] Forti L, Di Mauro S, Cramarossa MR, Filippucci S, Turchetti B, Buzzini P. Non-conventional yeasts whole cells as efficient biocatalysts for the production of flavors and fragrances. Molecules. 2015;**20**:10377-10398. DOI: 10.3390/molecules200610377

[9] Basso LC, Basso TO, Rocha SN. Ethanol production in Brazil: The industrial process and its impact on yeast fermentation. In: Bernardes MAS, editor. Biofuel Production-Recent Developments and Prospects. Rijeka: Intech; 2011. pp. 85-100. DOI: 10.5772/17047

[10] Taylor MP, Mulako I, Tuffin M, Cowan D. Understanding physiological responses to pre-treatment inhibitors in ethanologenic fermentations. Biotechnology Journal. 2012;**7**:1169-1181. DOI: 10.1002/biot.201100335

[11] Walker G. Lignocellulose to fuel alcohol: Current trends. Microbiology Today. 2013;**40**:150-153. Available from: http://www.sgm.ac.uk/en/publications/policy-docs.cfm/publication/biofuels/article/0D9268AC-7B45-434E-92A8443ADE2BFBF7 [Accessed: February 14, 2017]

[12] Urbaniec K. The evolution of evaporator stations in the beet-sugar industry. Journal of Food Engineering. 2004;**61**:505-508. DOI: 10.1016/S0260-8774(03)00218-8

[13] Meadows AL, Hawkins KM, Tsegaye Y, Antipov E, Kim Y, Raetz L, Dahl RH, Tai A, Mahatdejkul-Meadows T, Xu L, Zhao L, Dasika MS, Murarka A, Lenihan J, Eng D, Leng JS, Liu CL, Wenger JW, Jiang H, Chao L, Westfall P, Lai J, Ganesan S, Jackson P, Mans R, Platt D, Reeves CD, Saija PR, Wichmann G, Holmes VF, Benjamin K, Hill PW, Gardner TS, Tsong AE. Rewriting yeast central carbon metabolism for industrial isoprenoid production. Nature. 2016;**537**:694-697. DOI: 10.1038/nature19769

[14] Wu L, Moteki T, Gokhale AA, Flaherty DW, Toste FD. Production of fuels and chemicals from biomass: Condensation reactions and beyond. Chem. 2016;**1**:32-58. DOI: 10.1016/j.chempr.2016.05.002

[15] Jamai L, Ettayebi M. Bioethanol production process using the non-conventional yeast *Candida tropicalis*. In: Renewable and Sustainable Energy Conference (IRSEC); 2013. DOI: 10.1109/IRSEC.2013.6529710

[16] Mosier N, Wyman C, Dale B, Elander R, Lee YY, Holtzapple M, Ladisch M. Features of promising technologies for pretreatment of lignocellulosic biomass. Bioresource Technology. 2005;**96**:673-686. DOI: 10.1016/j.biortech.2004.06.025

[17] Wyman CE, Dale BE, Elander RT, Holtzapple M, Ladisch MR, Lee YY. Coordinated development of leading biomass pretreatment technologies. Bioresource Technology. 2005;**96**:1959-1966. DOI: 10.1016/j.biortech.2005.01.010

[18] Kang J, Kim D, Lee T. Hydrogen production and microbial diversity in sewage sludge fermentation preceded by heat and alkaline treatment. Bioresource Technology. 2012;**109**:239-243. DOI: 10.1016/j.biortech.2012.01.048

[19] Michalska K, Miazek K, Krzystek L, Ledakowicz S. Influence of pretreatment with Fenton's reagent on biogas production and methane yield from lignocellulosic biomass. Bioresource Technology. 2012;**119**:72-78. DOI: 10.1016/j.biortech.2012.05.105

[20] Harmsen PFH, Huijgen WJJ, Bermúdez López LM, Bakker RRC. Literature Review of Physical and Chemical Pretreatment Processes for Lignocellulosic Biomass. ECN-E-10-013; 2010. Available from: https://www.ecn.nl/docs/library/report/2010/e10013.pdf [Accessed: February 14, 2017]

[21] Rallis Ch. New tricks for an old-favorite model. Genome Biology. 2009;**10**:315. DOI: 10.1186/gb-2009-10-9-315

[22] Moysés DN, Reis VCB, Moreira de Almeida JR, Pepe de Moraes LM, Torre FAG. Xylose fermentation by *Saccharomyces cerevisiae*: Challenges and prospects. International Journal of Molecular Sciences. 2016;**17**:207. DOI: 10.3390/ijms17030207

[23] Wang C, Zhao J, Qiu C, Wang S, Shen Y, Du B, Ding Y, Ba X. Coutilization of D-glucose, D-xylose, and L-arabinose in *Saccharomyces cerevisiae* by coexpressing the metabolic pathways and evolutionary engineering. BioMed Research International. 2017, Article ID 5318232, 8 pages. DOI: 10.1155/2017/5318232

[24] Wisselink HW, Toirkens MJ, Wu O, Pronk JT, van Maris AJA. Novel evolutionary engineering approach for accelerated utilization of glucose, xylose, and arabinose mixtures by engineered *Saccharomyces cerevisiae* strains. Applied Microbiology and Biotechnology. 2009;**75**(4):907-914. DOI: 10.1128/AEM.02268-08

[25] Jeffries TW, Jin Y-S. Metabolic engineering for improved fermentation of pentoses by yeasts. Applied Microbiology and Biotechnology. 2004;**63**(5):495-509. DOI: 10.1007/s00253-003-1450-0

[26] Subtil T, Boles E. Competition between pentoses and glucose during uptake and catabolism in recombinant *Saccharomyces cerevisiae*. Biotechnology for Biofuels. 2012;**5**:14. Available from: http://www.biotechnologyforbiofuels.com/content/5/1/14 [Accessed: May 21, 2017]

[27] Jin Y-S, Laplaza JM, Jeffries TW. *Saccharomyces cerevisiae* engineered for xylose metabolism exhibits a respiratory response. Applied and Environmental Microbiology. 2004;**70**:6816-6825. DOI: 10.1128/AEM.70.11.6816-6825.2004

[28] Hahn-Hägerdal B, Karhumaa K, Fonseca C, Gorwa-Grauslund MF. Towards industrial pentose-fermenting yeast strains. Applied Microbiology and Biotechnology. 2007;**74**(5):937-953. DOI: 10.1007/s00253-006-0827-2

[29] Kuhad RC, Gupta R, Pal Khasa Y, Singh A, Zhang YHP. Bioethanol production from pentose sugars: Current status and future prospects. Renewable and Sustainable Energy Reviews. 2011;**15**(9):4950-4962. DOI: 10.1016/j.rser.2011.07.058

[30] Wolf K. Nonconventional Yeasts in Biotechnology: A Handbook. Springer-Verlag Berlin Heidelberg; 1996. DOI: 10.1007/978-3-642-79856-6

[31] Sibirny AA, Scheffers L. Thematic section 'Biochemistry, Genetics, Biotechnology and Ecology of Non-conventional Yeasts'. FEMS Yeast Research. 2002;**2**:293. Available from: http://onlinelibrary.wiley.com/doi/10.1111/j.1567-1364.2002.tb00097.x/pdf [Accessed: February 14, 2017]

[32] Kavšček M, Stražar M, Curk T, Natter K, Petrovič U. Yeast as a cell factory: Current state and perspectives. Microbial Cell Factories. 2015;**14**:94. DOI: 10.1186/s12934-015-0281-x

[33] Flores C-L, Rodriguez C, Petit T, Gancedo C. Carbohydrate and energy-yielding metabolism in non-conventional yeasts. FEMS Microbiology Reviews. 2000;**24**:507-529. DOI: 10.1111/j.1574-6976.2000.tb00553.x

[34] Souciet JL, Dujon B, Gaillardin C, Johnston M, Baret PV, Cliften P, Sherman DJ, Weissenbach J, Westhof E, Wincker P, Jubin C, Poulain J, Barbe V, Ségurens B, Artiguenave F, Anthouard V, Vacherie B, Val ME, Fulton RS, Minx P, Wilson R, Durrens P, Jean G, Marck C, Martin T, Nikolski M, Rolland T, Seret ML, Casarégola S, Despons L, Fairhead C, Fischer G, Lafontaine I, Leh V, Lemaire M, de Montigny J, Neuvéglise C, Thierry A, Blanc-Lenfle I, Bleykasten C, Diffels J, Fritsch E, Frangeul L, Goëffon A, Jauniaux N, Kachouri-Lafond R, Payen C, Potier S, Pribylova L, Ozanne C, Richard GF, Sacerdot C, Straub ML, Talla E. Comparative genomics of protoploid *Saccharomycetaceae*. Genome Research. 2009;**19**:1696-1709. DOI: 10.1101/gr.091546.109

[35] Ubeda J, Gil MM, Chiva R, Guillamon JM, Briones A. Biodiversity of non-*Saccharomyces* yeasts in distilleries of the La Mancha region (Spain). FEMS Yeast Research. 2014;**14**:663-673. DOI: 10.1111/1567-1364.12152

[36] Jeffries TW, Kurtzman C. Strain selection, taxonomy, and genetics of xylose-fermenting yeasts. Enzyme and Microbial Technology. 1994;(11):922-932. DOI: 10.1016/0141-0229(94)90001-9

[37] Jeffries TW, Jin Y-S. Ethanol and thermotolerance in the bioconversion of xylose by yeasts. Advances in Applied Microbiology. 2000;**47**:221-268. DOI: 10.1016/S0065-2164(00)47006-1

[38] Johnson EA. Biotechnology of non-*Saccharomyces* yeasts—The ascomycetes. Applied Microbiology and Biotechnology. 2013;**97**(2):503-517. DOI: 10.1007/s00253-012-4497-y

[39] Dulermo R, Brunel F, Dulermo T, Ledesma-Amaro R, Vion J, Trassaert M, Thomas S, Nicaud J-M, Leplat C. Using a vector pool containing variable-strength promoters to optimize protein production in *Yarrowia lipolytica*. Microbial Cell Factories. 2017;**16**:31. DOI: 10.1186/s12934-017-0647-3

[40] Madhavan A, Jose AA, Binod P, Sindhu R, Sukumaran RK, Pandey A, Castro GE. Synthetic biology and metabolic engineering approaches and its impact on non-conventional yeast and biofuel production. Frontiers in Energy Research. 2017;**5**:8. DOI: 10.3389/fenrg.2017.00008

[41] Kregiel D. Immobilization of yeast cells in alginate gels for ethanol production – Potentialities and limitations. Biotechnology and Food Science. 2005;**69**:59-66

[42] Djekrif DS, Gillmann L, Bennamoun L, Ait-Kaki A, Labbani K, Nouadri T, Meraihi Z. Amylolytic yeasts: Producers of α-amylase and pullulanase. International Journal of Life Sciences Scientific Research. 2016;**2**:339-354. Available from: http://ijlssr.com/currentissue/IJLSSR-1115-10-2015.pdf [Accessed: February 14, 2017]

[43] Kregiel D. Amylolytic activity of Kluyver-positive *Debaryomyces occidentalis* cells immobilized in foamed alginate gel. Journal of Microbiology, Biotechnology and Food Sciences. 2016;**5**:311-313. DOI: 10.15414/jmbfs.2016.5.4.311-313

[44] Kregiel D, Berlowska J, Ambroziak W. Growth and metabolic activity of conventional and non-conventional yeasts immobilized in foamed alginate. Enzyme and Microbial Technology. 2013;**53**:229-234. DOI: 10.1016/j.enzmictec.2013.05.0102013

[45] Negruta O, Csutak O, Stoica I, Rusu E, Vassu T. Methylotrophic yeasts: Diversity and methanol metabolism. Romanian Biotechnological Letters. 2010;**5**:5369-5375. Available from: http://www.rombio.eu/rbl4vol15/1%20Negruta.pdf [Accessed: February 14, 2017]

[46] Berlowska J, Dudkiewicz M, Kregiel D, Czyzowska A, Witonska I. Cell lysis induced by membrane-damaging detergent saponins from *Quillaja saponaria*. Enzyme and Microbial Technology. 2015;**75-76**:44-48. DOI: 10.1016/j.enzmictec.2015.04.007

[47] Berlowska J, Dudkiewicz-Kolodziejska M, Pawlikowska E, Pielech-Przybylska K, Balcerek M, Czyzowska A, Kregiel D. Utilisation of post-fermentation yeasts for yeast extract production by autolysis: The effect of yeast strain and saponin *Q. saponaria*. Journal of the Institute of Brewing. 2017. DOI: 10.1002/jib.438

[48] Martorell P, Stratford M, Steels H, Fernandez-Espinar MT, Querol A. Physiological characterization of spoilage strains of *Zygosaccharomyces bailii* and *Zygosaccharomyces rouxii* isolated from high sugar environments. International Journal of Food Microbiology. 2007;**114**:234-242. DOI: 10.1016/j.ijfoodmicro.2006.09.014

[49] Dujon B. Yeast evolutionary genomics. Nature Reviews Genetics. 2010;**11**:512-524. DOI: 10.1038/nrg2811

[50] Riley R, Haridas S, Wolfe KH, Lopes MR, Hittinger CT, Göker M, Salamov AA, Wisecaver JH, Long TM, Calvey CH, Aerts AL, Barry KW, Choi C, Clum A, Coughlan AY, Deshpande S, Douglass AP, Hanson SJ, Klenk H-P, LaButti KM, Lapidus A, Lindquist EA, Lipzen AM, Meier-Kolthoff JP, Ohm RA, Otillar RP, Pangilinan JL, Peng Y, Rokas A, Rosa CA, Scheuner C, Sibirny AA, Slot JC, Stielow JB, Sun H, Kurtzman CP, Blackwell M, Grigoriev IV, Jeffries TW. Comparative genomics of biotechnologically important yeasts. PNAS. 2016;**113**(35):9882-9887. DOI: 10.1073/pnas.1603941113

[51] Masneuf-Pomarede I, Bely M, Marullo P, Albertin W. The genetics of non-conventional wine yeasts: Current knowledge and future challenges. Frontiers in Microbiology. 2016;**6**:1563. DOI: 10.3389/fmicb.2015.01563

[52] Gunde-Cimerman N, Ramos J, Plemenitaš A. Halotolerant and halophilic fungi. Mycological Research. 2009;**113**(11):1231-1241. DOI: 10.1016/j.mycres.2009.09.002

[53] Hostinová E, Gašperík J. Yeast glucoamylases: Molecular-genetic and structural characterization. Biologia. 2010;**65**(4):559-568. DOI: 10.2478/s11756-010-0077-8

[54] Espinoza P, Barzana E, García-Garibay M, Gómez-Ruiz L. Evaluation of *Kluyveromyces marxianus* for the production of lactase simultaneously to pectinase or inulinase. Biotechnology Letters. 1992;**14**:1053-1058. DOI: 10.1007/BF010210581992

[55] Morrissey JP, Etschmann MMW, Schrader J, de Billerbec GM. Cell factory applications of the yeast *Kluyveromyces marxianus* for the biotechnological production of natural flavour and fragrance molecules. Yeast. 2015;**32**:3-16. DOI: 10.1002/yea.3054

[56] Watanabe J, Uehara K, Mogi Y, Suzuki K, Watanabe T, Yamazaki T. Improved transformation of the halo-tolerant yeast *Zygosaccharomyces rouxii* by electroporation. Bioscience, Biotechnology, and Biochemistry. 2010;**74**:1092-1094. DOI: 10.1271/bbb.90865

[57] Tao X, Zheng D, Liu T, Wang P, Zhao W, Zhu M, Jiang X, Zhao Y, Wu X. A novel strategy to construct yeast *Saccharomyces cerevisiae* strains for very high gravity fermentation. PLoS One. 2012;**7**:e31235. DOI: 10.1371/journal.pone.0031235

[58] Pais TM, Foulquie-Moreno MR, Hubmann G, Duitama J, Swinnen S, Goovaerts A, Yang Y, Dumortier F, Thevelein JM. Comparative polygenic analysis of maximal ethanol accumulation capacity and tolerance to high ethanol levels of cell proliferation in yeast. PLoS Genetics. 2013;**9**:e1003548. DOI: 10.1371/journal.pgen.1003548

[59] Erasmus DJ, van der Merwe GK, van Vuuren HJJ. Genome-wide expression analyses: Metabolic adaptation of *Saccharomyces cerevisiae* to high sugar stress. FEMS Yeast Research. 2003;**3**:375-399. DOI: 10.1016/S1567-1356(02)00203-9

[60] Wojda I, Alonso-Monge R, Bebelman JP, Mager WH, Siderius M. Response to high osmotic conditions and elevated temperature in *Saccharomyces cerevisiae* is controlled by intracellular glycerol and involves coordinate activity of MAP kinase pathways. Microbiology. 2003;**149**:1193-1204. DOI: 10.1099/mic.0.26110-0

[61] Pribylova L, de Montigny J, Sychrova H. Osmoresistant yeast *Zygosaccharomyces rouxii*: The two most studied wild-type strains (ATCC 2623 and ATCC 42981) differ in osmotolerance and glycerol metabolism. Yeast. 2007;**24**:171-180. DOI: 10.1002/yea.1470

[62] Leandro MJ, Sychrova H, Prista C, Loureiro-Dias MC. The osmotolerant fructophilic yeast *Zygosaccharomyces rouxii* employs two plasma-membrane fructose uptake systems belonging to a new family of yeast sugar transporters. Microbiology. 2011;**157**:601-608. DOI: 10.1099/mic.0.044446-0

[63] Aggarwal A, Mondal AK. *Debaryomyces hansenii*: An osmotolerant and halotolerant yeast. In: Satyanarayana T, Kunze G, editors. Yeast Biotechnology: Diversity and Applications. Springer Science + Business Media B.V.; 2009. pp. 65-84. DOI: 10.1007/978-1-4020-8292-4_4

[64] Hohman S. Osmotic stress signaling and osmoadaptation in yeasts. Microbiology and Molecular Biology Reviews. 2002;**66**(2):300-372. DOI: 10.1128/MMBR.66.2.300-372.2002

[65] Gonzalez-Hernandez JC, Cárdenas-Monroy CA, Peña A. Sodium and potassium transport in the halophilic yeast *Debaryomyces hansenii*. Yeast. 2004;**21**:403-412. DOI: 10.1002/yea.1108

[66] Prista C, Michán C, Miranda IM, Ramos J. The halotolerant *Debaryomyces hansenii*, the Cinderella of non-conventional yeasts. Yeast Primer. 2016;**33**(10):523-533. DOI: 10.1002/yea.3177

[67] Montiel V, Ramos J. Intracellular Na^+ and K^+ distribution in *Debaryomyces hansenii*. Cloning and expression in *Saccharomyces cerevisiae* of DhNHX1. FEMS Yeast Research. 2007;**7**:102-109. DOI: 10.1111/j.1567-1364.2006.00115.x

[68] Neves ML, Oliveira RP, Lucas CM. Metabolic flux response to salt-induced stress in the halotolerant yeast *Debaryomyces hansenii*. Microbiology. 1997;**143**:1133-1139. DOI: 10.1099/00221287-143-4-1133

[69] Pribylova L, Farkaš V, Slaninová I, de Montigny J, Sychrová H. Differences in osmotolerant and cell-wall properties of two *Zygosaccharomyces rouxii* strains. Folia Microbiologica. 2007;**52**:241-245. DOI: 10.1007/BF02931305

[70] Dakal TC, Solieri L, Giudici P. Adaptive response and tolerance to sugar and salt stress in the food yeast *Zygosaccharomyces rouxii*. International Journal of Food Microbiology. 2014;**185**:140-157. DOI: 10.1016/j.ijfoodmicro.2014.05.015

[71] Kumar S, Randhawa A, Ganesan K, Raghava GPS, Mondal AK. Draft genome sequence of salt-tolerant yeast *Debaryomyces hansenii* var. *hansenii* MTCC 234. Eukaryotic Cell. 2012;**11**(7):961-962. DOI: 10.1128/EC.00137-12

[72] Tabka MG, Gimbert I, Monod F, Sigoillot JC. Enzymatic saccharification of wheat straw for bioethanol production by a combined cellulose xylanase and feruloyl esterase treatment. Enzyme and Microbial Technology. 2006;**39**:897-902. DOI: 10.1016/j.enzmictec.2006.01.021

[73] Anderson PJ, McNeil K, Watson K. High-efficiency carbohydrate fermentation to ethanol at temperatures above 40°C by *Kluyveromyces marxianus* var. *marxianus* isolated from sugar mills. Applied and Environmental Microbiology. 1986;**51**:1314-1320. Available from: https://www.ncbi.nlm.nih.gov/pmc/articles/PMC239064/ [Accessed: February 14, 2017]

[74] Nonklang S, Abdel-Banat BMA, Cha-aim K, Moonjai N, Hoshida H, Limtong S, Yamada M, Akada R. High-temperature ethanol fermentation and transformation with linear DNA in the thermotolerant yeast *Kluyveromyces marxianus* DMKU3-1042. Applied and Environmental Microbiology. 2008;**74**:7514-7521. DOI: 10.1128/AEM.01854-08

[75] Abdel-Banat BMA, Hoshida H, Ano A, Nonklang S, Akada R. High-temperature fermentation: How can processes for ethanol production at high temperatures become superior to the traditional process using mesophilic yeast? Applied Microbiology and Biotechnology. 2010;**85**:861-867. DOI: 10.1007/s00253-009-2248-5

[76] Limtong S, Sringiew C, Yongmanitchai W. Production of fuel ethanol at high temperature from sugar juice by a newly isolated *Kluyveromyces marxianus*. Bioresource Technology. 2007;**98**:3367-3374. DOI: 10.1016/j.biortech.2006.10.044

[77] Fonseca GG, Heinzle E, Wittmann C, Gombert AK. The yeast *Kluyveromyces marxianus* and its biotechnological potential. Applied Microbiology and Biotechnology. 2008;**79**:339-354. DOI: 10.1007/s00253-008-1458-6

[78] Kurylenko OO, Ruchala J, Hryniv OB, Abbas CA, Dmytruk KV, Sibirny AA. Metabolic engineering and classical selection of the methylotropic thermotolerant yeast *Hansenula polymorpha* for improvement of high-temperature xylose alcoholic fermentation. Microbial Cell Factories. 2014;**13**:122-132. DOI: 10.1186/s12934-014-0122-3

[79] Gao L, Liu Y, Sun H, Li C, Zhao Z, Liu G. Advances in mechanisms and modifications for rendering yeast thermotolerance. Journal of Bioscience and Bioengineering. 2015;**121**(6):599-606. DOI: 10.1016/j.jbiosc.2015.11.002

[80] Mensonides FIC, Brul S, Hellingwerf KJ, Bakker BM, de Mattos MJT. A kinetic model of catabolic adaptation and protein reprofiling in *Saccharomyces cerevisiae* during temperature shifts. The FEBS Journal. 2014;**281**(3):825-841. DOI: 10.1111/febs.12649

[81] Schöffl F, Prändl R, Reindl A. Regulation of the heat-shock response. Plant Physiology. 1998;**117**(4):1135-1141. DOI: https://doi.org/10.1104/pp.117.4.1135

[82] Larkindale J, Hall JD, Knight MR, Vierling E. Heat stress phenotypes of *Arabidopsis* mutants implicate multiple signaling pathways in the acquisition of thermotolerance. Plant Physiology. 2005;**138**:882-897. DOI: 10.1104/pp.105.062257

[83] Crowe J. Trehalose as a 'chemical chaperone': Fact and fantasy. Advances in Experimental Medicine and Biology. 2007;**594**:143-158. DOI: 10.1007/978-0-387-39975-1_13

[84] Lin Y, Zhang W, Li C, Sakakibara K, Tanaka S, Kong H. Factors affecting ethanol fermentation using *Saccharomyces cerevisiae* BY4742. Biomass and Bioenergy. 2012;**47**:395-401. DOI. 10.1016/j.biombioe.2012.09.019

[85] Bokulich NA, Bamforth CW. The microbiology of malting and brewing. Microbiology and Molecular Biology Reviews. 2013;**77**(2):157-172. DOI: 10.1128/MMBR.00060-12

[86] Coote PJ, Jones MV, Seymour IJ, Rowe DL, Ferdinando DP, McArthur AJ, Cole MB. Activity of the plasma membrane H^+-ATPase is a key physiological determinant of thermotolerance in *Saccharomyces cerevisiae*. Microbiology. 1994;**140**:1881-1890. DOI: 10.1099/13500872-140-8-1881

[87] Shahsavarani H, Sugiyama M, Kaneko Y, Chuenchit B, Harashim S. Superior thermotolerance of *Saccharomyces cerevisiae* for efficient bioethanol fermentation can be achieved by overexpression of RSP5 ubiquitin ligase. Biotechnology Advances. 2012;**30**:1289-1300. DOI: 10.1016/j.biotechadv.2011.09.002

[88] Lecker SH, Goldberg AL, Mitch WE. Protein degradation by the ubiquitin–proteasome pathway in normal and disease states. Journal of American Society of Nephrology. 2006;**17**:1807-1819. DOI: 10.1681/ASN.2006010083

[89] Gibney PA, Lub C, Caudy AA, Hess DC, Botstein D. Yeast metabolic and signaling genes are required for heat-shock survival and have little overlap with the heat-induced genes. PNAS. 2013;**28**:E4393-E4402. DOI: 10.1073/pnas.1318100110/-/DCSupplemental

[90] Shui W, Xiong Y, Xiao W, Qi X, Zhang Y, Lin Y, Guo Y, Zhang Z, Wang Q, Ma Y. Understanding the mechanism of thermotolerance distinct from heat shock response through proteomic analysis of industrial strains of *Saccharomyces cerevisiae*. Molecular & Cellular Proteomics. 2015;**14**:1885-1897. DOI: 10.1074/mcp.M114.045781

[91] Jolly NP, Varela C, Pretorius IS. Not your ordinary yeast: Non-*Saccharomyces* yeasts in wine production uncovered. FEMS Yeast Research. 2014;**14**:215-237. DOI: 10.1111/1567-1364.12111

[92] Albergaria H, Arneborg N. Dominance of *Saccharomyces cerevisiae* in alcoholic fermentation processes: Role of physiological fitness and microbial interactions. Applied Microbiology and Biotechnology. 2016;**100**:2035-2046. DOI: 10.1007/s00253-015-7255-0

[93] van Dijken JP, Scheffers WA. Redox balances in the metabolism of sugars by yeasts. FEMS Microbiology Letters. 1986;**32**:199-224. DOI: 10.1111/j.15746968.1986.tb01194.x

[94] Barnett JA. A history of research on yeast 2: Louis Pasteur and his contemporaries, 1850-1880. Yeast. 2000;**16**:755-771. DOI: 10.1002/1097-0061(20000615)16:8<755::AID-YEA587> 3.0.CO;2-4

[95] Rodrigues F, Ludovico P, Leao C. Sugar metabolism in yeasts: An overview of aerobic and anaerobic glucose. In: Peter G, Rosa C, editors. Biodiversity and Ecophysiology of Yeasts. Berlin, Heidelberg: Springer-Verlag; 2006. pp. 101-121. DOI: 10.1007/3-540-30985-3_6

[96] Hagman A, Säll T, Piškur J. Analysis of the yeast short-term Crabtree effect and its origin. The FEBS Journal. 2014;**281**:4805-4814. DOI: 10.1111/febs.13019

[97] Henderson CM, Block DE. Examining the role of membrane lipid composition in determining the ethanol tolerance of *Saccharomyces cerevisiae*. Applied Environmental Microbiology. 2014;**80**:2966-2972. DOI: 10.1128/AEM.04151-13

[98] Galafassi S, Merico A, Pizza F, Hellborg L, Molinari F, Piskur J, Compagno C. *Dekkera/Brettanomyces* yeast for ethanol production from renewable sources under oxygen-limited and low-pH conditions. Journal of Industrial Microbiology and Biotechnology. 2011;**38**:1079-1088. DOI: 10.1007/s10295-010-0885-4

[99] Zha Y, Hossain AH, Tobola F, Sedee N, Havekes M, Punt PJ. *Pichia anomala* 29X: A resistant strain for lignocellulosic biomass hydrolysate fermentation. FEMS Yeast Research. 2013;**13**:609-617. DOI: 10.1111/1567-1364.12062

[100] Mukherjee V, Steensels J, Lievens B, Van de Voorde I, Verplaetse A, Aerts G, Willems KA, Thevelein JM, Verstrepen KJ, Ruyters S. Phenotypic evaluation of natural and industrial *Saccharomyces* yeasts for different traits desirable in industrial bioethanol production. Applied Microbiology and Biotechnology. 2014;**98**:9483-9498. DOI: 10.1007/s00253-014-6090-z

[101] Ruyters S, Mukherjee V, Verstrepen KJ, Thevelein JM, Willems KA, Lievens B. Assessing the potential of wild yeasts for bioethanol production. Journal of Industrial Microbiology and Biotechnology. 2015;**42**:39-48. DOI: 10.1007/s10295-014-1544-y

[102] Steensels J, Daenen L, Malcorps P, Derdelinckx G, Verachtert H, Verstrepen KJ. *Brettanomyces* yeasts—From spoilage organisms to valuable contributors to industrial fermentations. International Journal of Food Microbiology. 2015;**206**:24-38. DOI: 10.1016/j.ijfoodmicro.2015.04.005

[103] Radecka D, Mukherjee V, Mateo RQ, Stojiljkovic M, Foulquié-Moreno MR, Thevelein JM. Looking beyond *Saccharomyces*: The potential of non-conventional yeast species for desirable traits in bioethanol fermentation. FEMS Yeast Research. 2015;**15**:fov053. DOI: 10.1093/femsyr/fov053

[104] Schifferdecker AJ, Siurkus J, Andersen MR, Joerck-Ramberg D, Ling Z, Zhou N, Blevins JE, Sibirny AA, Piškur J, Ishchuk OP. Alcohol dehydrogenase gene ADH3 activates glucose alcoholic fermentation in genetically engineered *Dekkera bruxellensis* yeast. Applied Microbiology and Biotechnology. 2016;**100**:3219-3231. DOI: 10.1007/s00253-015-7266-x

[105] González Siso MI, Ramil E, Cerdán ME, Freire-Picos MA. Respirofermentative metabolism in *Kluyveromyces lactis*: Ethanol production and the Crabtree effect. Enzyme and Microbial Technology. 1996;**18**(8):585-591. DOI: 10.1016/0141-0229(95)00151-4

[106] Birch RM, Walker GM. Influence of magnesium ions on heat shock and ethanol stress responses of *Saccharomyces cerevisiae*. Enzyme and Microbial Technology. 2000;**26**:678-687. DOI: 10.1016/S0141-0229(00)00159-9

[107] Voordeckers K, Kominek J, Das A, Espinosa-Cantú A, De Maeyer D, Arslan A, Van Pee M, van der Zande E, Meert W, Yang Y, Zhu B, Marchal K, DeLuna A, Van Noort V, Jelier R, Verstrepen KJ. Adaptation to high ethanol reveals complex evolutionary pathways. PLoS Genetics. 2015;**11**:e1005635. DOI: 10.1371/journal.pgen.1005635

[108] Zha Y, Muilwij B, Coulier L, Punt PJ. Inhibitory compounds in lignocellulosic biomass hydrolysates during hydrolysate fermentation processes. Journal of Bioprocessing and Biotechniques. 2012;**2**:1. DOI: 10.4172/2155-9821.1000112

[109] Palmqvist E, Hahn-Hägerdal B. Fermentation of lignocellulosic hydrolysates. II: Inhibitors and mechanisms of inhibition. Bioresource Technology. 2000;**74**:25-33. DOI: 10.1016/S0960-8524(99)00161-3

[110] Almeida JRM, Modig T, Petersson A, Hähn-Hägerdal B, Lidén G, Gorwa-Grauslund MF. Increased tolerance and conversion of inhibitors in lignocellulosic hydrolysates by *Saccharomyces cerevsiae*. Journal of Chemical Technology and Biotechnology. 2007;**82**:340-349. DOI: 10.1002/jctb.1676

[111] Janzowski C, Glaab V, Samimi E, Schlatter J, Eisenbrand G. 5-Hydroxymethylfurfural: Assessment of mutagenicity, DNA-damaging potential and reactivity towards cellular glutathione. Food and Chemical Toxicology. 2000;**38**:801-809. DOI: 10.1016/S0278-6915(00)00070-3

[112] Lin FM, Qiao B, Yuan YJ. Comparative proteomic analysis of tolerance and adaptation of ethanologenic *Saccharomyces cerevisiae* to furfural, a lignocellulosic inhibitory compound. Applied and Environmental Microbiology. 2009;**75**:3765-3776. DOI: 10.1128/AEM.02594-08

[113] Martinez A, Rodriguez ME, Wells ML, York SW, Preston JF, Ingram LO. Detoxification of dilute acid hydrolysates of lignocellulose with lime. Biotechnology Progress. 2001;**17**:287-293. DOI: 10.1021/bp0001720

[114] Qian M, Tian S, Li X, Zhang J, Pan Y, Yang X. Ethanol production from dilute-acid softwood hydrolysate by co-culture. Applied Biochemistry and Biotechnology. 2006;**134**:273-283. DOI: 10.1385/ABAB:134:3:273

[115] Villarreal MLM, Prata AMR, Felipe MGA, Silva JBAE. Detoxification procedures of eucalyptus hemicellulose hydrolysate for xylitol production by *Candida guilliermondii*. Enzyme and Microbial Technology. 2006;**40**:17-24. DOI: 10.1016/j.enzmictec.2005.10.032

[116] Chandel AK, Kapoor RK, Singh A, Kuhad RC. Detoxification of sugarcane bagasse hydrolysate improves ethanol production by *Candida shehatea* NCIM 3501. Bioresource Technology. 2007;**98**:1947-1950. DOI: 10.1016/j.biortech.2006.07.047

[117] Tian S, Zhou G, Yan F, Yu Y, Yang X. Yeast strains for ethanol production from lignocellulosic hydrolysates during in situ detoxification. Biotechnology Advances. 2009;**27**:656-660. DOI: 10.1016/j.biotechadv.2009.04.008

[118] Mollapour M, Piper PW. Hog1 mitogen-activated protein kinase phosphorylation target the yeast Fps1 aquaglyceroporin for endocytosis, thereby rendering cell resistant to acetic acid. Molecular and Cellular Biology. 2007;**27**:6446-6456. DOI: 10.1128/MCB.02205-06

[119] Arneborg N, Jespersen L, Jakobsen M. Individual cells of *Saccharomyces cerevisiae* and *Zygosaccharomyces bailii* ehibit different short term intracellular pH responses to acetic acid. Archives of Microbiology. 2000;**174**:125-128. DOI: 10.1007/s002030000185

[120] Brett CL, Tukaye DN, Mukherjee S, Rao R. The yeast endosomal $Na^{+}(K^{+})/H^{+}$ exchanger Nhx1 regulates cellular pH to control vesicle trafficking. Molecular Biology of the Cell. 2005;**16**:1396-1405. DOI: 10.1091/mbc.E04-11-0999

[121] Pampulha ME, Loureiro-Dias MC. Activity of glycolytic enzymes of *Saccharomyces cerevisiae* in the presence of acetic acid. Applied Microbiology and Biotechnology. 1990;**34**:375-380. DOI: 10.1007/BF00170063

[122] Ding J, Huang X, Zhang L, Zhao N, Yang D, Zhang K. Tolerance and stress response to ethanol on the yeast *Saccharomyces cerevisiae*. Applied Microbiology and Biotechnology. 2009;**85**:253-263. DOI: 10.1007/s00253-009-2223-1

[123] Casey JR, Grinstein S, Orlowski J. Sensors and regulators of intracellular pH. Nature Reviews Molecular Cell Biology. 2010;**11**:50-61. DOI: 10.1038/nrm2820

[124] Cantarella M, Contarella L, Gallifuoco A, Spera A, Alfani F. Effect of inhibitors released during steam-explosion treatment of poplar wood on subsequent enzymatic hydrolysis and SSF. Biotechnology Progress. 2004;**20**:200-206. DOI: 10.1021/bp0257978

[125] Lindberg L, Santos AXS, Riezman H, Olsson L, Bettiga M. Lipidomic profiling of *Saccharomyces cerevisiae* and *Zygosaccharomyces bailii* reveals critical changes in lipid composition in response to acetic acid stress. PLoS One. 2013;**8**:e73936. DOI: 10.1371/journal.pone.0073936

[126] Ciani M, Comitini F, Mannazzu I, Domizio P. Controlled mixed culture fermentation: A new perspective on the use of non-*Saccharomyces* yeasts in winemaking. FEMS Yeast Research. 2010;**10**(2):123-133. DOI: 10.1111/j.1567-1364.2009.00579.x

[127] Ciani M, Comitini F. Yeast interactions in multi-starter wine fermentation. Current Opinion in Food Science. 2015;**1**:1-6. DOI: 10.1016/j.cofs.2014.07.001

[128] Yamaoka C, Kurita O, Kubo T. Improved ethanol tolerance of *Saccharomyces cerevisiae* in mixed cultures with *Kluyveromyces lactis* on high-sugar fermentation. Microbiological Research. 2014;**169**(12):907-914. DOI: 10.1016/j.micres.2014.04.007

[129] Verma G, Nigam P, Singha D, Chaudhary K. Bioconversion of starch to ethanol in a single-step process by coculture of amylolytic yeasts and *Saccharomyces cerevisiae* 21. Bioresource Technology. 2000;**72**(3):261-266. DOI: 10.1016/S0960-8524(99)00117-0

[130] Araque E, Parra C, Rodríguez M, Freer J, Jaime B. Selection of thermotolerant yeast strains *Saccharomyces cerevisiae* for bioethanol production. Enzyme and Microbial Technology. 2008;**43**(2):120-123. DOI: 10.1016/j.enzmictec.2008.02.007

[131] Borbala E, Balazs F, Mats G, Guido Z. Separate hydrolysis and co-fermentation for improved xylose utilization in integrated ethanol production from wheat meal and wheat straw. Biotechnology for Biofuels. 2012;**5**:12. DOI: 10.1186/1754-6834-5-12

[132] Koti S, Prashanthi S, Gentela J, Rao LV. Enhanced bioethanol production from wheat straw hemicellulose by mutant strains of pentose fermenting organisms *Pichia stipitis* and *Candida shehatae*. SpringerPlus. 2016;**5**:1545. DOI: 10.1186/s40064-016-3222-1

[133] Renault PE, Albertin W, Bely M. An innovative tool reveals interaction mechanisms among yeast populations under oenological conditions. Applied Microbiology and Biotechnology. 2013;**97**(9):4105-4119. DOI: 10.1007/s00253-012-4660-5

[134] Sharma RR, Singh D, Singh R. Biological control of postharvest diseases of fruit and vegetables by microbial antagonists: A review. Biological Control. 2009;**50**:205-221. DOI: 10.1016/j.biocontrol.2009.05.001

[135] Liu J, Sui Y, Wisniewski M, Droby S, Liu Y. Review: Utilization of antagonistic yeasts to manage postharvest fungal diseases of fruit. International Journal of Food Microbiology. 2013;**167**:153-160

[136] Kordowska-Wiater M. Yeasts as biological control agents for plants. Postepy Mikrobiologii. 2011;**50**:107-119. Available from: http://www.pm.microbiology.pl/web/archiwum/vol5022011107.pdf [Accessed: February 14, 2017]

[137] Parafati L, Vitale A, Restuccia C, Cirvilleri G. Biocontrol ability and action mechanism of food-isolated yeast strains against *Botrytis cinerea* causing post-harvest bunch rot of table grape. Food Microbiology. 2015;**47**:85-92. DOI: 10.1016/j.fm.2014.11.013

[138] Ruiz-Moyano S, Martín A, Villalobos MC, Calle A, Serradilla MJ, Córdoba MG, Hernández A. Yeasts isolated from figs (*Ficus carica* L.) as biocontrol agents of postharvest fruit diseases. Food Microbiology. 2016;**57**:45-53. DOI: 10.1016/j.fm.2016.01.003

[139] Sipiczki M. *Metschnikowia* strains isolated from botrytized grapes antagonize fungal and bacterial growth by iron depletion. Applied and Environmental Microbiology. 2006;**73**:6716-6724. DOI: 10.1128/AEM.01275-06

[140] Pawlikowska E, Kregiel D. Non-conventional yeast *Metschnikowia pulcherrima* and its application in biotechnology. Postepy Mikrobiologii. 2017 (in press)

[141] Oro L, Ciani M, Comitini F. Antimicrobial activity of *Metschnikowia pulcherrima* on wine yeasts. Journal of Applied Microbiology. 2014;**116**:1209-1217. DOI: 10.1111/jam.12446

[142] Michel M, Meier-Dörnberg T, Jacob F, Methner F-J, Wagner RS, Hutzler M. Review: Pure non-*Saccharomyces* starter cultures for beer fermentation with a focus on secondary metabolites and practical applications. Journal of the Institute of Brewing. 2016;**122**:569-587. DOI: 10.1002/jib.381

[143] Shekhawat K, Bauer FF, Setati ME. Impact of oxygenation on the performance of three non-*Saccharomyces* yeasts in co-fermentation with *Saccharomyces cerevisiae*. Applied Microbiology and Biotechnology. 2016;**101**:2479-2491. DOI: 10.1007/s00253-016-8001-y

Section 2

Old Issues - New Problems

Saccharomyces cerevisiae Peroxiredoxins in Biological Processes: Antioxidant Defense, Signal Transduction, Circadian Rhythm, and More

Melina C. Santos, Carlos A. Breyer,
Leonardo Schultz, Karen S. Romanello,
Anderson F. Cunha, Carlos A. Tairum Jr and
Marcos Antonio de Oliveira

Additional information is available at the end of the chapter

http://dx.doi.org/10.5772/intechopen.70401

Abstract

The yeast *Saccharomyces cerevisiae* is a model organism for biochemical and genetic studies, and several very important discoveries of fundamental biological processes have been conducted using this yeast as an experimental organism. An emerging concept, which is validated by several works using this organism, relies on the biological importance of oxidant species, specially the hydroperoxides. These molecules were formed during aerobic biological process and control several intracellular mechanisms such as a range of signaling pathways, cell cycle, programmed cell death, circadian rhythm, aging, and lifespan extension. Thereby, cellular homeostasis depends on a refined control of hydroperoxides levels and low-molecular-weight molecules in combination with antioxidant enzymes playing a role in this equilibrium. This proposal is focused on the *S. cerevisiae* peroxiredoxins and their role in peroxide decomposition, signal transduction, circadian clocks, and aging as model enzymes for the study and comprehension of these biological processes in living organisms, including humans.

Keywords: thiol-specific antioxidant protein, functional transitions, peroxidase, chaperone, overoxidation

1. Introduction

The use of *Saccharomyces cerevisiae* as a biological model in the field of oxidant species research represents a very important tool in an exciting area. Emerging concepts, validated by several works, revealed the importance of oxidant molecules in biological processes, especially the hydroperoxides [1, 2]. These molecules are formed during several aerobic biological processes and, in adequate levels, are involved in a number of intracellular mechanisms, such as redox signaling pathways related to cell cycle progression, programmed cell death, circadian rhythm, aging, and lifespan extension [2–7]. However, the accumulation of these molecules can be harmful to the cells [3, 4]. In fact, highly deleterious radical species can be generated from hydroperoxides, such as hydroxyl radical (•OH), that is generated from hydrogen peroxide (H_2O_2) through Fenton and Harber-Weiss reactions. The •OH is able to oxidize carbohydrates, lipids, proteins, and DNA, being extremely toxic to cells. Thereby, cellular homeostasis depends on a refined control of hydroperoxides levels, and this role is played by both low-molecular-weight molecules, such as glutathione and ascorbic acid, as well as by antioxidant enzymes such as glutathione peroxidases (Gpxs), catalases (Cats), and peroxiredoxins (Prxs) [2, 4]. The latter ones have been subject of intense studies since works involving kinetic approaches indicate that the Prxs decompose more than 90% of cellular hydroperoxides [8, 9]. Additionally, to exert their biological functions, several Prxs are able to perform amazing structural switches, revealing an intricate puzzle among protein structure and function [10–12].

The first Prx described was a cytosolic enzyme identified in *S. cerevisiae* and received the name of "thiol-specific antioxidant protein 1" (Tsa1) [13]. Subsequently, a second homologue cytosolic isoform, named Tsa2, was identified and characterized. Currently, there are five Prx isoforms identified in this yeast. In mammals, there are six isoforms described, and as in other organisms, they are located in several cellular environments as cytosol, nucleus, peroxisome, mitochondria, endoplasmic reticulum, and even in the nucleus [14, 15]. Furthermore, these proteins are very abundant. For example, in *S. cerevisiae*, they can reach ~0.9% of total soluble proteins and can represent one of ten most expressed enzymes in bacteria and in mammal cells [16]. In human erythrocytes, PrxII is the third most abundant protein, only losing in concentration for globins and carbonic anhydrase, and its level is modulated during cell differentiation [17].

Besides the widespread cellular distribution and abundance, Prx stands out due to their highly efficient ability to decompose a wide variety of hydroperoxides (H_2O_2, nitrite peroxide, lipid peroxides, among others), with second order rates reaching ~10^6–10^8 M^{-1} s^{-1}) [18–21]. These characteristics place the Prx as one of the main modulators of hydroperoxides levels and, consequently, of the cellular processes mediated by them. The Prx enzymes are able to decompose hydroperoxides without any prosthetic group, but using only a highly reactive cysteine residue named peroxidatic cysteine (C_P) [5, 22]. All the Prxs described to date present a conserved motif containing the C_P (PXXXT/SXXC$_P$), which is oxidized to cysteine sulfenic acid (C_P-SOH) after hydroperoxide reduction [10]. This enzyme family is very heterogeneous, and different classifications have been proposed; the most currently used one subdivides these proteins in three large subclasses,1-Cys Prx, typical 2-Cys Prx, and atypical 2-Cys Prx, based in the number of cysteines involved in catalytic cycle and structural aspects (**Figure 1**). The 1-Cys Prxs are homodimeric proteins that present only one cysteine residue, the C_P, involved in hydroperoxide catalysis. 2-Cys Prxs may be monomeric (in the case of some atypical 2-Cys Prx) or homodimeric proteins and present a second cysteine residue, named resolving cysteine (C_R), which condenses with C_P forming a disulfide bond as final product during the catalytic cycle. In typical 2-Cys Prx, the disulfide is intermolecular (e.g., between different monomers), while in atypical 2-Cys Prx, the disulfide is intramolecular (in the same monomer) [23].

Among the different Prx subclasses, the typical 2-Cys Prxs are the best studied, and, from this point on, our focus will be on this Prx subclass. After oxidation, the disulfide bond of the typical 2-Cys Prx is frequently reduced by the low-molecular-weight (~11 kDa) enzyme thioredoxin (Trx). The oxidized Trx is reduced by thioredoxin reductase (TrxR), which uses electrons from nicotinamide adenine dinucleotide phosphate (NADPH) *via* a flavin adenine dinucleotide (FAD) molecule. Together, Trx, TrxR, and NADPH are named thioredoxin system (Trx system) [21]. It is important to mention that all electron exchanges between the proteins are performed using catalytic cysteines [25] (**Figure 2**).

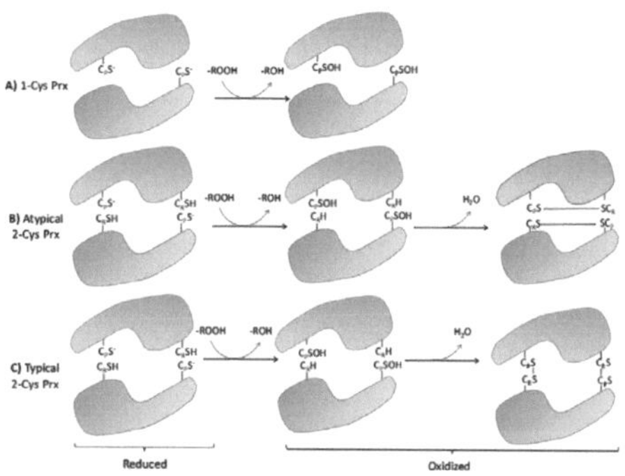

Figure 1. Prx subclasses in reduced and oxidized states. For all enzymes, the first step of the catalytic cycle is represented by the attack of the C_P-S^- over the O—O from hydroperoxide forming cysteine sulfenic acid (C_P-SOH) and releasing R—OH. (A) 1-Cys are dimeric enzymes containing only the peroxidatic cysteine, which is stable in oxidized state (C_P-SOH). (B) In the atypical 2-Cys Prx, the oxidized cysteine (C_P-SOH) formed after hydroperoxide decomposition condenses with the C_R-SH from the same monomer forming an intramolecular disulfide bond. (C) In typical 2-Cys Prx, the C_P-SOH condenses with C_R from the adjacent monomer forming an intermolecular disulfide.

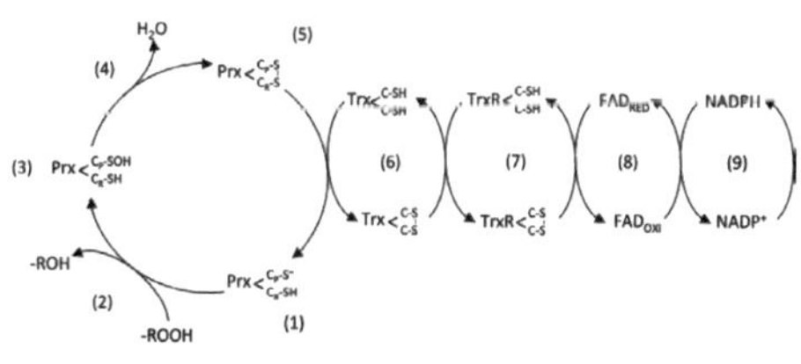

Figure 2. Hydroperoxide reduction steps by typical 2-Cys Prx and Trx system. The Prx C_P in thiolate form (**1**) attacks the hydroperoxide (**2**), releasing a water molecule in the case of H_2O_2 reduction, or an alcohol when the substrate is an organic hydroperoxide (the "R" represents the hydroperoxide radical). C_P is oxidized to cysteine sulfenic acid (**3**), releases a water molecule (**4**) and condenses with C_R forming an intermolecular disulfide (**5**), which is reduced by the enzyme Trx (**6**). Trx disulfide is reduced by the cysteines from TrxR enzyme (**7**) using electrons from NADPH (**9**) *via* a FAD molecule (**8**).

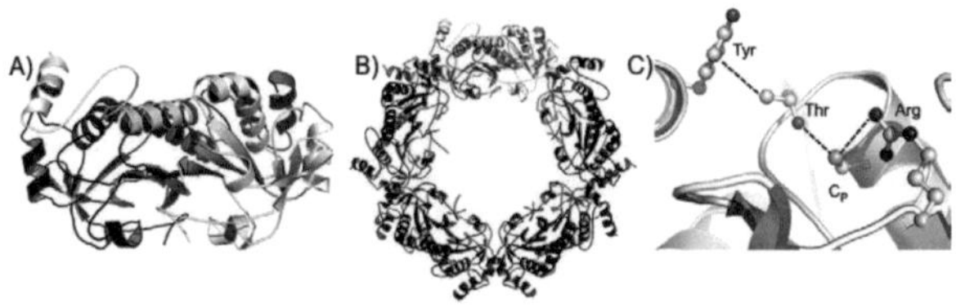

Figure 3. Quaternary structures of the typical 2-Cys Prx. (A) The yeast Tsa1 homodimer is represented in cartoon. (B) α2(5) decamer formed by the association of five homodimers. (C) Microenvironment of the CP in the active site. The Thr and Arg residues are involved in the thiolate (S−) stabilization. Additionally, the Thr residue is able to perform a CH-π interaction with the C atoms of a Tyr ring from the adjacent dimer. The proteins are represented in cartoon, and catalytic triad and the Tyr residue are represented in ball and stick. Figures were generated using the S. cerevisiae Tsa1 crystallographic coordinates (PDB: 3SBC) and Pymol software (http://www.pymol.org/).

Despite that the basic functional unit of the typical 2-Cys Prx is represented by a α(2) homodimer, studies using the Tsa1 and Tsa2 isoforms from *S. cerevisiae* revealed that this oligomeric state presents low peroxidase activity, and the highest reactivity of the typical 2-Cys Prx is reached when these proteins are found in a ring-shaped α2(5) decamers (association of five homodimers; **Figure 3A**). It is believed that the alternation between these two quaternary structures is responsible for the modulation of their peroxidase activity and may be involved in signal transduction (**Figure 3B**) [10]. Additionally, the typical 2-Cys Prx enzymes may also present other oligomeric states that will be discussed posteriorly.

The high reactivity of Prx over hydroperoxides is related to the maintenance of C_P in thiolate form (C_P-S^-), suitable for catalysis as a consequence of the microenvironment of the active site. The C_P thiolate is stabilized by polar interactions with a threonine (or a serine, in some cases) and an arginine residue (**Figure 3C**). These three residues (Thr, C_P, and Arg) are named catalytic triad and are widely conserved among all Prxs described to date [10]. During catalysis, a guanidine group of the Arg residue is able to perform a hydrogen bond with the proximal oxygen (O) of the hydroperoxide, allowing the nucleophilic attack of the C_P over the hydroperoxide [24]. The Oγ from Thr, in turn, would act as an acceptor of the hydrogen bond with the distal O from hydroperoxide, aiding the positioning of the molecule in a productive way to catalysis [24].

Typical 2-Cys Prxs, such as *S. cerevisiae* Tsa1 and Tsa2, are still able to perform additional structural and functional switches acting as peroxide sensors, molecular chaperones and are involved in several hydroperoxide-dependent signal transduction pathways, as it will be discussed further [20, 21]. Tsa1 and Tsa2 are also evolutionarily related to human PrxI and PrxII. In fact, Tsa1 presents 67% of identity and 77% of similarity with human PrxII, while Tsa2 presents 60% of identity and 76% of similarity with human PrxI, which places these proteins as important models to the study of the human Prx and the biological processes related to them.

2. Redox cycle and structural transitions

During the redox cycle, some typical 2-Cys Prxs are able to transit between different oligomeric species: α2(5) decamers (reduced enzyme) and α2 dimers (disulfide oxidized protein). Aiming to understand the details of the catalytic cycle and structural transitions, we have

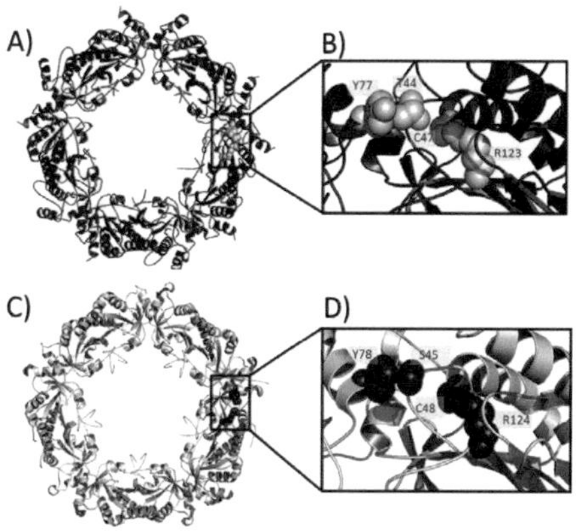

Figure 4. Tsa1 and Tsa2 interactions at dimer-dimer interface. Cartoon representations of quaternary structures of the Tsa1 and Tsa2. (A) Representation of theTsa1 decamer in cartoon. (B) Interaction between Thr44 and Tyr77 in Tsa1 decamer interface. (C) Tsa2 decamer in cartoon. (D) Interaction between Ser45 and Tyr78 in the Tsa2 decamer interface. In (B) and (D) the atoms are represented by spheres and colored as follow: O = red, C = yellow, N = blue. The gures were generated using the S. cerevisiae Tsa1 (PDB: 3SBC) and Tsa2 (PDB: 5DVB) coordinates and the Pymol software (http://www.pymol.org/).

determined the crystallographic structure of Tsa1 [21]. In fact, the analysis of the structure revealed an interaction of the Thr from the active site motif, at the dimer-dimer interface of the decamer. Recently, using different methodological approaches as site-directed mutagenesis, biochemical approaches, size exclusion chromatography, and structural analysis, we have demonstrated that a slight difference in the PXXXT(S)XXC_P is involved in decamers to dimers transitions [10]. While Tsa1 possess a Thr residue embedded in the conserved motif, in Tsa2, the Thr is naturally substituted by a Ser (**Figure 4**). In fact, the Tsa1 enzyme, containing Thr residue, transits between dimers (oxidized form) and decamers (reduced enzyme), but the Ser-containing enzyme Tsa2 is not able to dissociate in dimers. Indeed, the rearrangements as consequence of the redox states in the Tsa1 may cause hysteric hindrance of the Thr Oγ with the Tyr aromatic ring of the adjacent monomer, causing the decamer dissociation. Since Tsa2 presents a Ser residue, the hysteric clash probably is avoided. These characteristics may indicate an additional regulation of Prx quaternary structure, which may have implications in biological processes.

3. Prx overoxidation: structural and functional implications

During the typical 2-Cys Prx catalytic cycle under high levels of hydroperoxides, before disulfide formation, C_P-SOH can be attacked by another hydroperoxide molecule and becomes overoxidized to cysteine sulfinic acid (C_P-SO_2H) or sulfonic acid (C_P-SO_3H). The C_P overoxidation is related to spectacular functional and structural switch in typical 2-Cys Prx. As mentioned before, when the typical 2-Cys Prx are in reduced state (C_P-S^-), these proteins are decamers, but when are oxidized in disulfide, they can be dimers and/or decamers and are able to act as peroxidases (**Figure 5A** and **B**) [9, 11]. However, when the C_P is overoxidized, these enzymes are able to promote an intense oligomerization to form high-molecular-weight (HMW) spherical complexes (**Figure 5C**), with the concomitant inactivation of the peroxidase

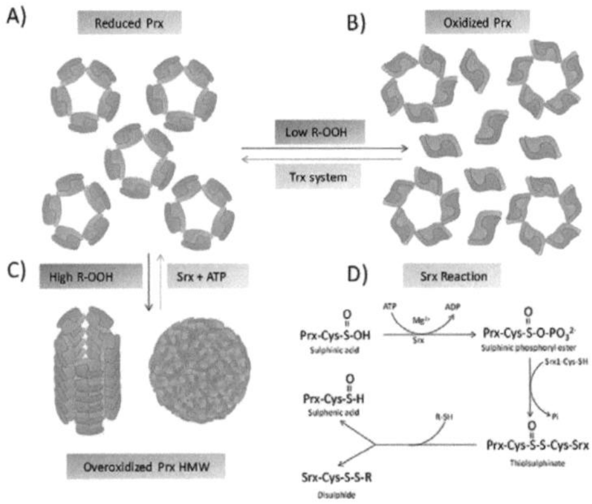

Figure 5. Typical 2-Cys Prx overoxidation and high-molecular-weight complex formation. The typical 2-Cys Prx in reduced form are presented as α2(5) decamers (A). In low concentrations of hydroperoxides, the C_P is oxidized in C_P-SOH, and the intermolecular disulfide is formed with C_R. The disulfide formation, in some cases, is able to destabilize the decamers, forming a mixture of decamers and dimers (B). The oxidized form is reduced by Trx system. When the typical 2-Cys Prx are challenged with high concentrations of hydroperoxides, the C_P can be overoxidized to C_P-SO_2H. The C_P overoxidation promotes the HMW complexes formation which presents chaperone properties (C). The C_P-SO_2H can be reduced by sulfiredoxin, in ATP and Mg^{2+} dependent manner (D).

activity. The HMW complexes formation was first reported in *S. cerevisiae* Tsa1 and Tsa2 by Jang and coworkers [11], and, posteriorly, very similar complexes were described to the human homologues typical 2-Cys Prxs (PrxI and PrxII) [26]. Using transmission electron microscopy (TEM), it was demonstrated that complexes are represented by heterogeneous spherical structures, which can reach 1 GDa, and biochemical approaches revealed that the complexes present an extraordinary chaperone holdase activity [9, 26]. Later on, similar spherical and another type of HWM complexes, represented by the stacking of several decamers (**Figure 5C**), were described for the plant chloroplastic 2-Cys Prxs, cyanobacterial Anabaena PCC7120 2-Cys Prx, among others [27–30]. The structural differences between the HMW complexes are not well understood to date.

A very important point relies on the fact that the Prx overoxidized species cannot be reduced by the Trx system, but some studies revealed that overoxidized typical 2-Cys Prx species could be regenerated to the reduced form *in vivo* [31]. Posteriorly, it was identified in *S. cerevisiae*, and after in human and other species, an enzyme named sulfiredoxin (Srx) which is able to reduce the C_P-SO_2H in a ATP and Mg^{2+} dependent reaction, but not C_P-SO_3H, suggesting that this oxidation state is refractive to reduction [31]. Curiously, the C_P-SO_2H reduction rates by Srx are very slow when compared to the disulfide reduction by Trx (~2 M^{-1} s^{-1} and ~10^6 M^{-1} s^{-1}, respectively) [32]. The biochemical steps of the C_P-SO_2H reduction by Srx are represented in **Figure 5D**. It is important to highlight that the Srx was identified in several eukaryotes, but few prokaryotes possess this enzyme, which may be an evolutionary sophistication of the 2-Cys Prx redox cycle [33]. In fact, to the majority of the prokaryotes, no homologous Srx gene was detected in their genomes, and the typical 2-Cys Prxs are much more resistant to overoxidation. Moreover, an additional classification can be done based on the C_P overoxidation susceptibility, and the 2-Cys Prx can be classified as sensitive or robust.

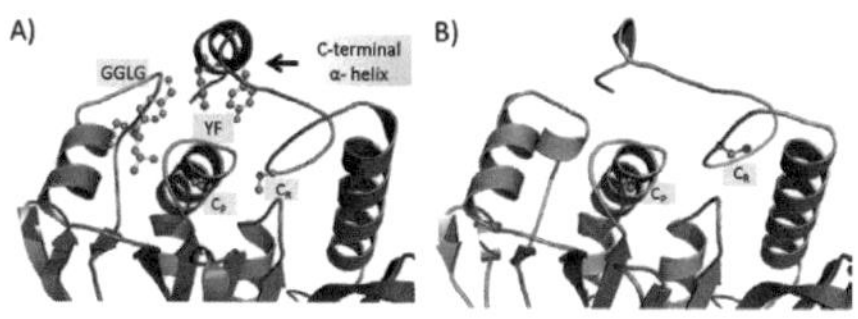

Figure 6. Structural comparison of the sensitive S. cerevisiae Tsa1 and the robust S. typhimurium AhpC. The comparison of Tsa1 (A) and AhpC (B) structures reveal the presence of the GGLG and YF motifs typically found in eukaryotes. The structures are represented in cartoon and the structural motifs as well the CP and CR are represented in ball and sticks. The figures were generated using the S. cerevisiae Tsa1 (PDB: 3SBC) and S. typhimurium AhpC (PDB: 4MA9) coordinates in Pymol software (http://www.pymol.org/).

The sensitive enzymes are present in eukaryotes and in some cyanobacteria, and the robust 2-Cys Prxs are exclusive to prokaryotes [34, 35]. The structural analyses of sensitive versus robust 2-Cys Prx revealed the presence of two motifs in the sensitive 2-Cys Prx. One is an insertion with conserved Gly-Gly-Leu-Gly, denominated GGLG motif (**Figure 6A**), and the other is an additional α-helix in C-terminal extension with a conserved Tyr-Phe sequence, the YF motif, both involved in C_P overoxidation susceptibility. **Figure 6** shows the comparison of *S. cerevisiae* Tsa1, a sensitive typical 2-Cys Prx, and *Salmonella typhimurium* (AhpC), a robust enzyme [34]. This difference is associated with important effects in redox cell signaling transduction and will be detailed in the next topics.

4. Typical 2-Cys Prx roles in redox signal transduction pathways

Increasing evidence shows the involvement of the typical 2-Cys Prx with the redox signal transduction pathways. Several antioxidant coding genes are activated by the transcriptional regulator activator protein 1 (AP1) which is considered as the major transcriptional activator of the antioxidant proteins in eukaryotes. It has been shown that the translocation of the homologue factor in budding yeast (YAP1) from cytosol to the nucleus may be controlled by 2-Cys Prx indirectly by the modulation of the cytosolic hydroperoxide levels [36]. In mammals, the PrxII is able to perform a physical interaction with the transcription factor STAT 3 (signal transducer and activator of transcription 3), which is able to activate the transcription of several genes involved in cell growth and apoptosis [37]. The authors demonstrated that PrxII can form mixed disulfides through C_P and cysteine residues of the DNA binding and trans-activating domains from STAT3, attenuating its transcriptional activity. Although the direct interaction of the typical 2-Cys Prx with target proteins is still an emerging area, this work reveals that the Prx may be an ultrasensitive hydroperoxide sensor that can form transient disulfides with unknown target proteins, which may have implications in biological processes. Additionally, the mammal PrxI can bind to several proteins including the tumor suppressor phosphatase and tensin homolog (PTEN), protecting it against suppression of its lipid phosphatase activity, which occurs under oxidative stress. On the other hand, PTEN deficiency causes decrease of PrxI, PrxII, PrxV, and PrxVI, suggesting that the Prxs and PTEN act together to maintain cellular antioxidant levels and suppress cancer-promoting pathways, such as the PI3K-Akt pathway [38].

Despite the importance of the physical interaction between typical 2-Cys Prx and biological targets, the indirect role in the regulation of the cell-signaling redox pathways is dependent of an intricate balance between peroxiredoxin, thioredoxin, and sulfiredoxin levels and their redox state. As an example, in yeast, the number of Tsa1 molecules per cell is estimated in 378,000 in aerobic conditions (log phase, SD medium), while its reductants represented by Trx and Srx molecules are much lower (~13,000 and 538 molecules/cell, respectively) [16]. In the case of Trx enzymes, additionally to the Prx reduction, these enzymes are involved in several biological processes as deoxyribonucleotide synthesis, repair of oxidatively damaged proteins, protein folding, sulfur metabolism, and activation of transcription factors among others [16]. The importance of Tsa1 reduction by Trx in redox signaling promoted by hydrogen peroxide may be significant in the cells since it produces oxidized Trx, and many signal transduction pathways are only activated by the reduced Trx enzyme [1, 39]. Because the oxidation of Trx by hydroperoxidesis is negligible, Prxs may act as a catalyst of this reaction in the cells [21].

The typical 2-Cys Prx inactivation by the C_P overoxidation combined with the low rates of the reduction of the C_P-SO_2H by sulfiredoxin (~2 M^{-1} s^{-1}) [32] is able to enhance levels of the reduced Trx to participate of other biological processes. In fact, it has been shown that the C_P overoxidation of the typical 2-Cys Prx from *Schizosaccharomyces pombe* (Tpx1) enhance the levels of the reduced Trx and allow the repair of damaged proteins increasing cell survival [40]. Accordingly, in mammals, only the reduced form of Trx is able to bind to the apoptosis signal regulating kinase (Ask-1), inhibiting the apoptosis, thus revealing a redox-dependent signal transduction pathway, which is induced by Trx oxidation [41]. Also in mammals, the activation of the nuclear factor kappa light chain enhancer of activated β cells (NF-kβ), a transcription factor that plays a central role regulating pathways of immune and inflammatory processes [42], is dependent on the reduction of a cysteine residue by Trx [43]. Additionally, Trx is involved in the reduction and activation of several transcription factors as the tumor-suppressor p53, the glucocorticoid and estrogen receptors, and c-Fos/c-Jun complexes [39].

Finally, the direct modulation of peroxides levels is an important role of the 2-Cys Prx enzymes in cell growth. It has been shown that PrxI and PrxII can eliminate the intracellular hydrogen peroxide generated by the receptors stimulation. Overexpression of PrxI and PrxII in culture cells dramatically reduces the intracellular hydrogen peroxide levels generated in response to platelet-derived growth factor (PDGF), epidermal growth factor (EGF), tirotropin (TSH), and TNF-related apoptosis inducing ligands (TRAIL). Furthermore, it has been shown that the expression of these proteins also led to a block of NF-κβ activity, which is induced by the extracellular addition of H_2O_2 or tumor necrosis factor α (TNF-α) [44]. It has also been shown that PrxII regulates different MAP kinases. Under stimulation of TNF, in which the activity of PrxII was blocked or partially abolished (knockout and RNAi), the activity of JNK and P38 MAP kinase was increased [45]. Due to the involvement of PrxI and PrxII in cell growth events, several studies have demonstrated that these isoforms have elevated levels in distinct types of cancers in different organs and tissues such as esophagus, pancreas, thyroid, lung, and breast [44–46]. The high expression of PrxI/PrxII is also associated to a more aggressive phenotype of cancer cells resistant to chemotherapy and radiotherapy [44–46].

Some authors argued that the typical 2-Cys Prx enzymes maintain hydrogen peroxide in appropriate levels to cell growth but not to apoptosis. However, Liu et al. [47] showed that the neoplastic cells of acute myeloid leukemia treated with an inhibitor of the PrxI and PrxII peroxidase activity demonstrated that the accumulation of intracellular H_2O_2 is related to the activation of the ERK1 and ERK2 (extracellular signal regulatory kinases). The kinases activation leads to an increase in the expression of the CCAAT-enhancer-binding proteins β (C/EBPβ). This condition resulted in cell differentiation and consequent tumor regression [47, 48], showing additional complexity of the neoplastic processes with the involvement of the Prx. Since there is a notable resemblance between human and yeasts typical 2-Cys Prx as also other proteins of these pathways, yeasts may be used to explore these mechanisms.

5. Prx structural switch and circadian rhythm

The circadian rhythm is a fundamental process considered to be a feature of almost all living cells. The organisms are able to exhibit cycles in their metabolism, physiology, and behavior, even when isolated from external stimuli, maintaining a 24-h period [49]. However, the molecular mechanisms which drive the circadian rhythm are not simple to elucidate, since the already identified clock genes and proteins are not very conserved across phylogenetic kingdoms [49–53]. A common model for molecular mechanism has been described for all organisms which had their circadian rhythm investigated, named transcription-translation feedback loop (TTFL) [49]. However, the TTFL components are not shared between organisms, suggesting independent evolutionary processes. Additionally, it was showed that nontranscriptional mechanisms are sufficient to sustain circadian timekeeping in the eukaryotic lineage, although they normally function in conjunction with transcriptional components [51].

Recently, it has been demonstrated that in human erythrocytes, a cell type without transcriptional activity, the PrxI and PrxII exhibit an approximate 24-h rhythm according to the C_P overoxidation. This characteristic is shared with several organisms, including *S. cerevisiae*, indicating that the typical 2-Cys Prxs constitute a universal rhythmic biomarker [52]. To reach this conclusion, the authors performed immunoblotting analyses using a Prx C_P-$SOH_{2/3}$ antibody and showed that 2-Cys Prx proteins from organisms of different domains have been oscillated to overoxidized Prx species, in constant conditions, exhibiting a circadian oscillation, probably reflecting an endogenous rhythm in the generation of reactive oxygen species (ROS; **Figure 7**) [51, 54–56]. Because all living organisms possess typical 2-Cys Prx enzymes that present remarkable conservation of the active site, the same antibody was able to detect overoxidized typical 2-Cys Prx in mice, fungi, plants, bacteria, and archaea. This indicates that the circadian clock mechanism is likely conserved across phylogenetic domains [54].

Yeast Tsa1 and Tsa2 isoforms exhibit relationship with the shorter period yeast respiratory oscillations, a cell-autonomous, temperature-compensated rhythm in oxygen consumption that synchronizes spontaneously when cells are grown at high density in aerobic, nutrient-limited, continuous culture [52]. Additionally, the yeast respiratory oscillation cycle shares

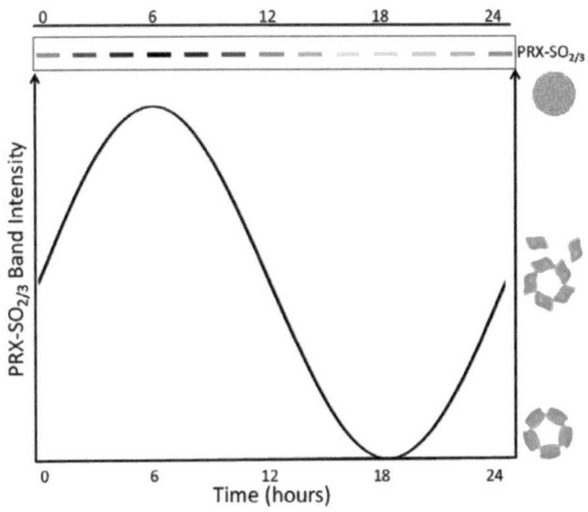

Figure 7. Redox circadian cycle of typical 2-Cys Prx. The circadian cycle of 2-Cys Prx could be detected by PRX-$SOH_{2/3}$ immunoblot. Western blot representation shows that overoxidized Prx has a circadian rhythm (upper part of the figure), and, consequently, the oligomeric state follows the redox state from Prx, alternating between dimers and decamers, in oxidative and reduced states, respectively, and HMW formation in overoxidized species (represented in the right side of the figure).

key features with the clock in mammalian cells, which may contribute to the elucidation about the origins of biological timekeeping [52].

Finally, it has been determined that the deregulation of the circadian rhythm is related to aging and genetic diseases [57]. Curiously, it has also been demonstrated that aging is related to the accumulation of the 2-Cys Prx overoxidized species in mammals [58]. Recently, a study involving the overoxidation of Tsa1 revealed that the chaperone activity detected in overoxidized species may be attributed to the association of this protein with the heat shock proteins Hsp70/Hsp104, revealing a pathway where the hydrogen peroxide is directly related to the aging process [12]. The authors also showed that the disaggregation process of the protein is dependent of Srx. Another study demonstrated that the presence of a mutant allele of Tsa1 resulted in accelerated aging in yeast [59]. One of the reasons for the involvement of these enzymes in the senescence process resides in the increase of the level of C_P overoxidation in Prx over time, even in the absence of oxidative stress [6]. In fact, this process also involves the caloric restriction, a well-known intervention that extends life span [60]. The caloric restriction elevates the level of Srx, which is responsible to reduce the hyperoxidized Tsa1, the inhibition of Tsa1 causes a profound genome instability, like chromosomal rearrangements and recombination, therefore increasing aging process [6, 61].

Another situation in which Prxs are involved is in the telomere length homeostasis [62]. The telomere dysfunction causes cellular senescence due to DNA damage [63]. The yeast mutant with *tsa1* gene deleted displayed reduction of telomere lengthening, which was not observed in conditions of low-oxidative exposure, probably due to the role of Tsa1 in hydroperoxide decomposition, avoiding DNA damage [62]. The understanding of the aging process and its implications in yeast can be used to extrapolate to higher eukaryotes. In fact, even in erythrocytes, the 2-Cys Prxs are related to the aging process. PrxII also has the ability to associate with the erythrocyte cell membrane through the N-terminal cytoplasmic domain of band 3 protein, after which PrxII undergoes a conformational change that does not entail the loss of its peroxidase function. This association may indicate a potential role of this Prx in the protection of membrane lipids against oxidative damage increasing life span [64]. Accordingly, a study carried out using mouse erythrocytes showed that the levels of overoxidized PrxII are due to autoxidation of hemoglobin and to PrxII degradation by the 20S proteasome. Approximately 1% of PrxII-SO_2H is degraded daily, leading to progressive loss of this enzyme which is directly related with the erythrocyte senescence [65]. Additionally, the aging process is directly related the genome instability. This instability is maintained in part by Prx action, and it is also involved in some diseases like cancer. Together, the data presented here reveal a cross talk of the 2-Cys Prx C_P overoxidation in circadian clocks, aging, and lifespan.

6. Methodologies to detect different redox species of the typical 2-Cys Prx

Several methodologies such as transmission electron microscopy (TEM), cryo-electron microscopy (Cryo-EM), size exclusion chromatography (SEC), mass spectrometry (MS), two-dimensional gel electrophoresis (2DGE), nonreducing SDS PAGE, and immunoblotting can be used to explore directly or indirectly the redox state of typical 2-Cys Prx [11, 66, 67]. However, for some experimental procedures, high-cost equipment and/or complex experimental procedures are necessary. Among these techniques, the nonreducing SDS PAGE, immunoblotting, and SEC are very good and cost-effective procedures, since no expensive equipment or complicated protocols are required. In this topic, these techniques and some experimental procedures will be discussed.

To access the formation of HMW complexes of purified 2-Cys Prx samples, the size exclusion chromatography (SEC) is the best choice. This methodology was used in the pioneer work performed by Jang and coworkers [11] using Tsa1 and Tsa2. In our lab, we performed a similar assay, using Tsa1, Trx system, and high concentration of cumene hydroperoxide (CHP) to promote the HMW complexes formation. Using SEC methodology, it is possible to separate several molecular species with mass range from ~45 kDa, correspondent to a dimer, followed by a 200-kDa peak representing the decameric species, several oligomeric intermediates, and a prominent specie with more than 1000 kDa (**Figure 8**). These results are in accordance with structural analyses performed by transmission electron microscopy (TEM) by Jang and coworkers [11], using negative stain. These authors analyzed different fractions separated by SEC, and their results revealed three distinct oligomeric configurations: large spherical shaped particles, heterogeneous spherical particles, and ring-shaped structures, as represented in **Figure 8**. Currently, the cryo-electron microscopy development has provided pronounced advances to resolve complex protein structures in high resolution, such as the human Prx [67].

To reduced, oxidized, and overoxidized species from purified proteins samples *in vitro*, a simple nonreducing SDS PAGE (without DTT or another reductant) can be used to detect different redox species of the enzymes. In fact, the reduce Prx decamers and HMW complexes are held together by weak molecular forces as hydrophobic, van der Walls, and polar interactions, that are disrupted in SDS PAGE. As an example, on a nonreducing gel containing SDS,

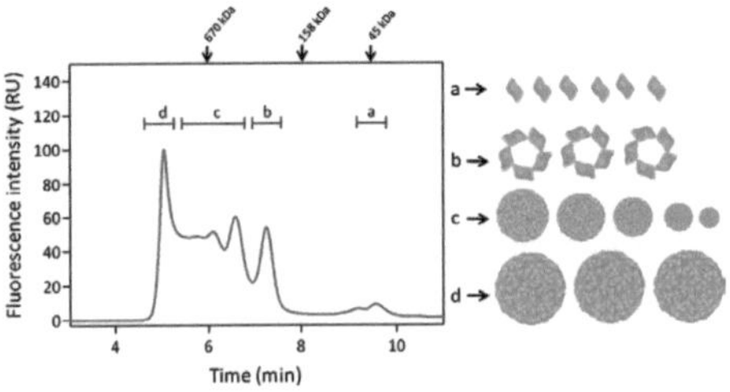

Figure 8. Overoxidized Tsa1 complexes analyzed by SEC. Tsa1 HMW species formation was analyzed by size-exclusion chromatography. The assay was performed overnight at 4°C in Hepes-NaOH 50mM (pH 7.4), DTPA 100 μM, sodium azide 1 mM, NADPH 1 mM, S. cerevisiae Prx 43.6 μM; S. cerevisiae Trx1 1 μM; S. cerevisiae TrxR1 0.3 μM and CHP 10 mM. The reaction was injected into the system containing a BioSep-SEC-S3000 column, eluted at a ow rate of 1 ml min−1 and monitored by tryptophan uorescence (excitation, 280 nm; emission, 340 nm). The elution pro le of the molecular standards thyroglobulin (bovine) (670 kDa), γ-globulin (bovine) (158 kDa), and ovalbumin (chicken) (44 kDa) were used to identify the 2-Cys Prx oligomers.

the Tsa1 is detected as a monomer (~25 kDa). The oxidized form is detected as ~50 kDa bands (dimer) as a consequence of the intermolecular disulfide bond that is formed between the C_P and C_R that is not disrupted in the gel. The overoxidized forms (Cys_P-SO_2H or Cys_P-SO_3H) can also be visualized as monomers, since the disulfide bond formation is not achievable [68, 69] (**Figure 9A** and **B**).

Figure 9B shows the SDS-PAGE result of *in vitro* procedure to perform Tsa1 overoxidation using growing concentrations of organic hydroperoxide (*t*-BOOH) and the Trx system (see the legend for detail). In this example, it is possible to verify the presence of the Tsa1 overoxidized species when high concentrations of *t*-BOOH were used (**Figure 9B**, upper panel). In **Figure 9C**, it is represented the probable quaternary structure present in the correspondent lane of the gel. In low concentrations of hydroperoxides, there are, predominantly, reduced Tsa1 in decameric

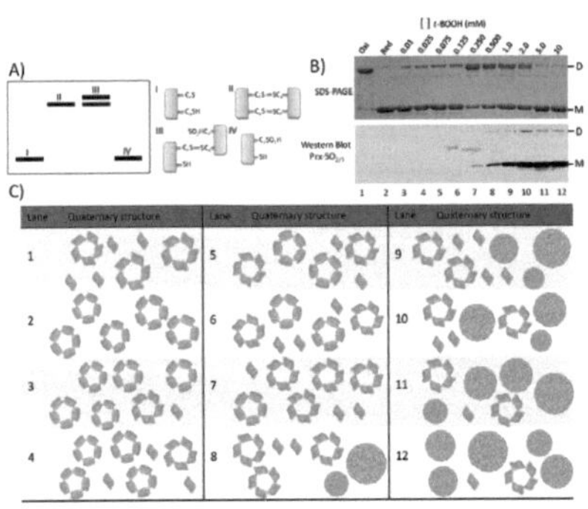

Figure 9. Redox state analyses by nonreducing SDS-PAGE and immunoblotting of the typical 2-Cys Prx. Diagram of the different 2-Cys Prx redox species in SDS-PAGE in nonreducing conditions by monomer or dimer formation (A). The Tsa1 overoxidation can be followed by SDS-PAGE in nonreducing conditions using *in vitro* approaches with Trx system in growing concentrations of hydroperoxides (B). In the example, the reaction was performed in a final volume of 50 μl at 30°C in Hepes-NaOH 50 mM (pH 7.0), DTPA 100 μM, sodium azide 1 mM, NADPH 150 μM, *S. cerevisiae* Tsa1 9.3 μM; *S. cerevisiae* Trx1 1 μM; *S. cerevisiae* TrxR1 0.3 μM and growing concentrations of *t*-BOOH (0.01, 0.025, 0.05, 0.75, 0.125, 0.25, 0.5, 1, 5, and 10 mM).The C_P overoxidized species can be observed in higher concentrations of *t*-BOOH. The oxidative state of Tsa1 induces the quaternary structural changes (C). The numbers in C demonstrate the possible structure of Tsa1 in different oxidative states (dimers, reduced decamers, oxidized decamers, and HMW complexes).

form, which is disrupted in SDS PAGE. At intermediate hydroperoxide concentrations, disulfide oxidized forms are detected, being represented by dimers and weak decamers (**Figure 9C**), that are detected as dimers in SDS PAGE. However, in high concentration of hydroperoxide, C_P overoxidation and HMW structure formations that, in the gel, are detected as monomers occur [11] (**Figure 9B**). To confirm the redox state result, a western blot analysis can be performed using an anti-$SO_{2/3}$ anti-body (**Figure 9B**, lower panel) [31, 70]. Additionally, the immunoblot technique can be used in nonpurified samples, as cell extracts, since antibodies to typical 2-Cys Prx are commercially available by several suppliers. Moreover, as mentioned before, the antibodies can be used in different species since the enzymes possess remarkable conservation [54].

7. Conclusions

S. cerevisiae is continuously used as a model organism by several researchers, being associated with significant advances in life sciences. In this chapter, we exposed several discoveries related with the role of the yeast Prx as a model in several studies related to hydroperoxide detoxification and signaling, and how these characteristics influence physiological processes like circadian rhythm and aging and diseases like cancer. All these features are related to the redox state of Prx and amazing functional and structural switches and the cross talk with different pathways that are regulated by hydroperoxide levels. Additionally, we present some practical approaches which can be easily implemented to Prx studies, like nonredox SDS-PAGE, size exclusion chromatography, and transmission electron microscopy. We believe that the use of these techniques may facilitate the study of these intricate enzymes for those interested in joining to this exciting research area.

Acknowledgements

We acknowledge the financial support from the Fundação de Amparo à Pesquisa do Estado de São Paulo grants 07/50930-3, 16/10130-7, and 13/07937-8 (Redox Processes in Biomedicine [REDOXOMA]).

Author details

Melina C. Santos[1,2], Carlos A. Breyer[1], Leonardo Schultz[1], Karen S. Romanello[3], Anderson F. Cunha[3], Carlos A. Tairum Jr[1,4] and Marcos Antonio de Oliveira[1]*

*Address all correspondence to: mao@clp.unesp.br

1 Biosciences Institute, São Paulo State University, São Vicente, Brazil

2 Biochemical-Pharmaceutical Technology Department, University of São Paulo, Brazil

3 Genetics and Evolution Department, Federal University of São Carlos, Brazil

4 Institute of Biosciences of the University of São Paulo, Brazil

References

[1] Netto LES, Antunes F. The roles of peroxiredoxin and thioredoxin in hydrogen peroxide sensing and in signal transduction. Molecules and Cells. 2016;**39**:65-71. DOI: 10.14348/molcells.2016.2349

[2] Rhee SG. Overview on peroxiredoxin. Molecules and Cells. 2016;**39**:1-5. DOI: 10.14348/molcells.2016.2368

[3] Finkel T. Oxidant signals and oxidative stress. Current Opinion in Cell Biology. 2003;**15**:247-254. DOI: 10.1016/S0955-0674(03)00002-4

[4] Halliwell B, Gutteridge JMC. Free Radicals in Biology and Medicine. 5th ed. Oxford: Oxford University Press; 2015.

[5] Kang SW, Rhee SG, Chang T-S, Jeong W, Choi MH. 2-Cys peroxiredoxin function in intracellular signal transduction: Therapeutic implications. Trends in Molecular Medicine. 2005;**11**:571-578. DOI: 10.1016/j.molmed.2005.10.006

[6] Molin M, Yang J, Hanzén S, Toledano MB, Labarre J, Nyström T. Life span extension and H(2)O(2) resistance elicited by caloric restriction require the peroxiredoxin Tsa1 in *Saccharomyces cerevisiae*. Molecular Cell. 2011;**43**:823-833. DOI: 10.1016/j.molcel.2011.07.027

[7] Vivancos AP, Castillo EA, Biteau B, Nicot C, Ayté J, Toledano MB, Hidalgo E. A cysteine-sulfinic acid in peroxiredoxin regulates H_2O_2-sensing by the antioxidant Pap1 pathway. Proceedings of the National Academy of Sciences of the United States of America. 2005;**102**:8875-8880. DOI: 10.1073/pnas.0503251102

[8] Adimora NJ, Jones DP, Kemp ML. A model of redox kinetics implicates the thiol proteome in cellular hydrogen peroxide responses. Antioxidants and Redox Signaling. 2010;13:731-743. DOI: 10.1089/ars.2009.2968

[9] Perkins A, Nelson KJ, Parsonage D, Poole LB, Karplus PA. Peroxiredoxins: Guardians against oxidative stress and modulators of peroxide signaling. Trends in Biochemical Sciences. 2015;**40**:435-445. DOI: 10.1016/j.tibs.2015.05.001

[10] Tairum CA, Santos MC, Breyer CA, Geyer RR, Nieves CJ, Portillo-Ledesma S, Ferrer-Sueta G, Toledo JC, Toyama MH, Augusto O, Netto LES, de Oliveira MA. Catalytic Thr or Ser residue modulates structural switches in 2-cys peroxiredoxin by distinct mechanisms. Scientific Reports. 2016;**6**:33133. DOI: 10.1038/srep33133

[11] Jang HH, Lee KO, Chi YH, Jung BG, Park SK, Park JH, Lee JR, Lee SS, Moon JC, Yun JW, Choi YO, Kim WY, Kang JS, Cheong GW, Yun DJ, Rhee SG, Cho MJ, Lee SY. Two enzymes in one: Two yeast peroxiredoxins display oxidative stress-dependent switching from a peroxidase to a molecular chaperone function. Cell. 2004;**117**:625-635. DOI: 10.1016/j.cell.2004.05.002

[12] Hanzén S, Vielfort K, Yang J, Roger F, Andersson V, Zamarbide-Forés S, Andersson R, Malm L, Palais G, Biteau B, Liu B, Toledano MBB, Molin M, Nyström T. Lifespan control by redox-dependent recruitment of chaperones to misfolded proteins. Cell. 2016;**166**:140-151. DOI: 10.1016/j.cell.2016.05.006

[13] Chae HZ, Kim IH, Kim K, Rhee SG. Cloning, sequencing, and mutation of thiol-specific antioxidant gene of *Saccharomyces cerevisiae*. Journal of Biological Chemistry. 1993;**268**:16815-16821

[14] Park SG, Cha MK, Jeong W, Kim IH. Distinct physiological functions of thiol peroxidase isoenzymes in *Saccharomyces cerevisiae*. Journal of Biological Chemistry. 2000;**275**: 5723-5732

[15] Pedrajas JR, Miranda-Vizuete A, Javanmardy N, Gustafsson JA, Spyrou G. Mitochondria of *Saccharomyces cerevisiae* contain one-conserved cysteine type peroxiredoxin with thioredoxin peroxidase activity. Journal of Biological Chemistry. 2000;**275**:16296-16301. DOI: 10.1074/jbc.275.21.16296

[16] Ghaemmaghami S, Huh W, Bower K, Howson RW, Belle A, Dephoure N, O'Shea EK, Weissman JS. Global analysis of protein expression in yeast. Nature. 2003;**425**:737-741. DOI: 10.1038/nature02046

[17] Low FM, Hampton MB, Winterbourn CC. Peroxiredoxin 2 and peroxide metabolism in the erythrocyte. Antioxidants and Redox Signaling. 2008;**10**:1621-1630 DOI: 10.1089/ars.2008.2081

[18] Horta BB, Oliveira MA, Discola KF, Cussiol JRR, Netto LES. Structural and biochemical characterization of peroxiredoxin Qbeta from Xylella fastidiosa: Catalytic mechanism and high reactivity. Journal of Biological Chemistry. 2010;**285**:16051-16065. DOI: 10.1074/jbc.M109.094839

[19] Toledo JC, Audi R, Ogusucu R, Monteiro G, Netto LES, Augusto O. Horseradish peroxidase compound i as a tool to investigate reactive protein-cysteine residues: From quantification to kinetics. Free Radical Biology and Medicine. 2011;**50**:1032-1038. DOI: 10.1016/j.freeradbiomed.2011.02.020

[20] Reyes AM, Hugo M, Trostchansky A, Capece L, Radi R, Trujillo M. Oxidizing substrate specificity of *Mycobacterium tuberculosis* alkyl hydroperoxide reductase E: Kinetics and mechanisms of oxidation and overoxidation. Free Radical Biology and Medicine. 2011;**51**:464-473. DOI: 10.1016/j.freeradbiomed.2011.04.023

[21] Tairum CA, de Oliveira MA, Horta BB, Zara FJ, Netto LES. Disulfide biochemistry in 2-cys peroxiredoxin: Requirement of Glu50 and Arg146 for the reduction of yeast Tsa1 by thioredoxin. Journal of Molecular Biology. 2012;**424**:28-41. DOI: 10.1016/j.jmb.2012.09.008

[22] Flohé L, Harris JR. Introduction. History of the peroxiredoxins and topical perspectives. Subcellular Biochemistry. 2007;**44**:1-25

[23] Flohé L, Toppo S, Cozza G, Ursini F. A comparison of thiol peroxidase mechanisms. Antioxidants and Redox Signaling. 2011;**15**:763-780. DOI: 10.1089/ars.2010.3397

[24] Hall A, Parsonage D, Poole LB, Karplus PA. Structural evidence that peroxiredoxin catalytic power is based on transition-state stabilization. Journal of Molecular Biology. 2010;**402**:194-209. DOI: 10.1016/j.jmb.2010.07.022

[25] Netto LES, de Oliveira MA, Tairum CA, da Silva Neto JF. Conferring specificity in redox pathways by enzymatic thiol/disulfide exchange reactions. Free Radical Research. 2016;**50**:206-245. DOI: 10.3109/10715762.2015.1120864

[26] Moon JC, Hah Y-S, Kim WY, Jung BG, Jang HH, Lee JR, Kim SY, Lee YM, Jeon MG, Kim CW, Cho MJ, Lee SY. Oxidative stress-dependent structural and functional switching of a human 2-Cys peroxiredoxin isotype II that enhances HeLa cell resistance to H_2O_2-induced cell death. Journal of Biological Chemistry. 2005;**280**:28775-28784. DOI: 10.1074/jbc.M505362200

[27] Toledano MB, Huang B. Microbial 2-Cys peroxiredoxins: Insights into their complex physiological roles. Molecules and Cells. 2016;**39**:31-39. DOI: 10.14348/molcells.2016.2326

[28] Banerjee M, Chakravarty D, Ballal A. Redox-dependent chaperone/peroxidase function of 2-Cys-Prx from the cyanobacterium Anabaena PCC7120: Role in oxidative stress tolerance. BMC Plant Biology. 2015;**15**:60. DOI: 10.1186/s12870-015-0444-2

[29] König J, Galliardt H, Jütte P, Schäper S, Dittmann L, Dietz K-J. The conformational bases for the two functionalities of 2-cysteine peroxiredoxins as peroxidase and chaperone. Journal of Experimental Botany. 2013;**64**:3483-3497. DOI: 10.1093/jxb/ert184

[30] Pan Y, Jin J-H, Yu Y, Wang J. Significant enhancement of hPrx1 chaperone activity through lysine acetylation. Chembiochem. 2014;**15**:1773-1776. DOI: 10.1002/cbic.201402164

[31] Biteau B, Labarre J, Toledano MB. ATP-dependent reduction of cysteine-sulphinic acid by *S. cerevisiae* sulphiredoxin. Nature. 2003;**425**:980-984. DOI: 10.1038/nature02075

[32] Mishra M, Jiang H, Wu L, Chawsheen HA, Wei Q. The sulfiredoxin-peroxiredoxin (Srx-Prx) axis in cell signal transduction and cancer development. Cancer Letters. 2015;**366**:150-159. DOI: 10.1016/j.canlet.2015.07.002

[33] Wood Z, Poole LB, Karplus PA. Peroxiredoxin evolution and the regulation of hydrogen peroxide signaling. Science. 2003;**300**:650-653. DOI: 10.1126/science.1080405

[34] Wood ZA, Schro E, Harris JR, Poole LB. Structure, mechanism and regulation of peroxiredoxins. Trends in Biochemical Sciences. 2003;**28**:00003-00008

[35] Pascual MB, Mata-cabana A, Florencio FJ, Lindahl M, Cejudo FJ. Overoxidation of 2-Cys peroxiredoxin in prokaryotes. Journal of Biological Chemistry. 2010;**285**:34485-34492. DOI: 10.1074/jbc.M110.160465

[36] Tachibana T, Okazaki S, Murayama A, Naganuma A, Nomoto A, Kuge S. A major peroxiredoxin-induced activation of Yap1 transcription factor is mediated by reduction-sensitive disulfide bonds and reveals a low level of transcriptional activation. Journal of Biological Chemistry. 2009;**284**:4464-4472. DOI: 10.1074/jbc.M807583200

[37] Sobotta MC, Liou W, Stöcker S, Talwar D, Oehler M, Ruppert T, Scharf AND, Dick TP. Peroxiredoxin-2 and STAT3 form a redox relay for H_2O_2 signaling. Nature Chemical Biology. 2015;**11**:64-70. DOI: 10.1038/nchembio.1695

[38] Cao J, Schulte J, Knight A, Leslie NR, Zagozdzon A, Bronson R, Manevich Y, Beeson C, Neumann CA. Prdx1 inhibits tumorigenesis via regulating PTEN/AKT activity. EMBO Journal. 2009;**28**:1505-1517. DOI: 10.1038/emboj.2009.101

[39] Berndt C, Lillig CH, Holmgren A. Thiol-based mechanisms of the thioredoxin and glutaredoxin systems: Implications for diseases in the cardiovascular system. American Journal of Physiology—Heart and Circulatory Physiology. 2007;**292**:H1227-H1236. DOI: 10.1152/ajpheart.01162.2006

[40] Day AM, Brown JD, Taylor SR, Rand JD, Morgan BA, Veal EA. Article Inactivation of a peroxiredoxin by hydrogen peroxide is critical for thioredoxin-mediated repair of oxidized proteins and cell survival. Molecular Cell. 2012;**45**:398-408. DOI: 10.1016/j.molcel.2011.11.027

[41] Saitoh M, Nishitoh H, Fujii M, Takeda K, Tobiume K, Sawada Y, Kawabata M, Miyazono K, Ichijo H. Mammalian thioredoxin is a direct inhibitor of apoptosis signal-regulating kinase (ASK) 1. EMBO Journal. 1998;**17**:2596-2606. DOI: 10.1093/emboj/17.9.2596

[42] Napetschnig J, Wu H. Molecular basis of NF-κB signaling. Annual Review of Biophysics. 2013;**42**:443-468. DOI: 10.1146/annurev-biophys-083012-130338

[43] Matthews JR, Wakasugi N, Virelizier JL, Yodoi J, Hay RT. Thioredoxin regulates the DNA binding activity of NF-kappa B by reduction of a disulphide bond involving cysteine 62. Nucleic Acids Research. 1992;**20**:3821-3830

[44] Lee JY, Jung HJ, Song IS, Williams MS, Choi C, Rhee SG, Kim J, Kang SW. Protective role of cytosolic 2-cys peroxiredoxin in the TNF-alpha-induced apoptotic death of human cancer cells. Free Radical Biology and Medicine. 2009;**47**:1162-1171. DOI: 10.1016/j.freeradbiomed.2009.07.027

[45] Choi MH, Lee IK, Kim GW, Kim BU, Han Y-H, Yu D-Y, Park HS, Kim KY, Lee JS, Choi C, Bae YS, Lee BI, Rhee SG, Kang SW. Regulation of PDGF signalling and vascular remodelling by peroxiredoxin II. Nature. 2005;**435**:347-353. DOI: 10.1038/nature03587

[46] Miki H, Funato Y. Regulation of intracellular signalling through cysteine oxidation by reactive oxygen species. Journal of Biochemistry. 2012;**151**:255-261. DOI: 10.1093/jb/mvs006

[47] Liu C-X, Yin Q-Q, Zhou H-C, Wu Y-L, Pu J-X, Xia L, Liu W, Huang X, Jiang T, Wu M-X, He L-C, Zhao Y-X, Wang X-L, Xiao W-L, Chen H-Z, Zhao Q, Zhou A-W, Wang L-S, Sun H-D, Chen G-Q. Adenanthin targets peroxiredoxin I and II to induce differentiation of leukemic cells. Nature Chemical Biology. 2012;**8**:486-493. DOI: 10.1038/nchembio.935

[48] Liu C-X, Zhou H-C, Yin Q-Q, Wu Y-L, Chen G-Q. Targeting peroxiredoxins against leukemia. Experimental Cell Research. 2013;**319**:170-176. DOI: 10.1016/j.yexcr.2012.06.013

[49] Dunlap JC. Molecular bases for circadian clocks. Cell. 1999;**96**:271-290. DOI: 10.1016/S0092-8674(00)80566-8

[50] Woelfle MA, Ouyang Y, Phanvijhitsiri K, Johnson CH. The adaptive value of circadian clocks: An experimental assessment in Cyanobacteria. Current Biology. 2004;**14**:1481-1486. DOI: 10.1016/j.cub.2004.08.023

[51] O'Neill JS, van Ooijen G, Dixon LE, Troein C, Corellou F, Bouget F-Y, Reddy AB, Millar AJ. Circadian rhythms persist without transcription in a eukaryote. Nature. 2011;**469**: 554-558. DOI: 10.1038/nature09654

[52] Causton HC, Feeney KA, Ziegler CA, O'Neill JS. Metabolic cycles in yeast share features conserved among circadian rhythms. Current Biology. 2015;**25**:1056-1062. DOI: 10.1016/j.cub.2015.02.035

[53] Wijnen H, Young MW. Interplay of circadian clocks and metabolic rhythms. Annual Review of Genetics. 2006;**40**:409-448. DOI: 10.1146/annurev.genet.40.110405.090603

[54] Edgar RS, Green EW, Zhao Y, van Ooijen G, Olmedo M, Qin X, Xu Y, Pan M, Valekunja UK, Feeney KA, Maywood ES, Hastings MH, Baliga NS, Merrow M, Millar AJ, Johnson CH, Kyriacou CP, O'Neill JS, Reddy AB. Peroxiredoxins are conserved markers of circadian rhythms. Nature. 2012;**485**:459-464. DOI: 10.1038/nature11088

[55] Hoyle NP, O'Neill JS. Oxidation-reduction cycles of peroxiredoxin proteins and non-transcriptional aspects of timekeeping. Biochemistry. 2015;**54**:184-193. DOI: 10.1021/bi5008386

[56] O'Neill JS, Reddy AB. Circadian clocks in human red blood cells. Nature. 2011;**469**:498-503. DOI: 10.1038/nature09702

[57] Fonseca Costa SS, Ripperger JA. Impact of the circadian clock on the aging process. Frontiers in Neurology. 2015;**6**:1-5. DOI: 10.3389/fneur.2015.00043

[58] Nyström T, Yang J, Molin M. Peroxiredoxins, gerontogenes linking aging to genome instability and cancer. Genes and Development. 2012;**26**:2001-2008. DOI: 10.1101/gad.200006.112

[59] Timmermann B, Jarolim S, Rußmayer H, Kerick M, Michel S, Krüger A, Bluemlein K, Laun P, Grillari J, Lehrach H, Breitenbach M, Ralser M. A new dominant peroxiredoxin allele identified by whole-genome re-sequencing of random mutagenized yeast causes oxidant-resistance and premature aging. Aging (Albany NY). 2010;**2**:475-486. DOI: 10.18632/aging.100187

[60] Kaeberlein M, Hu D, Kerr EO, Tsuchiya M, Westman EA, Dang N, Fields S, Kennedy BK. Increased life span due to calorie restriction in respiratory-deficient yeast. PLoS Genetics. 2005;**1**:e69. DOI: 10.1371/journal.pgen.0010069

[61] Huang M-E, Kolodner RD. A biological network in *Saccharomyces cerevisiae* prevents the deleterious effects of endogenous oxidative DNA damage. Molecular Cell. 2005;**17**:709-720. DOI: 10.1016/j.molcel.2005.02.008

[62] Lu J, Vallabhaneni H, Yin J, Liu Y. Deletion of the major peroxiredoxin Tsa1 alters telomere length homeostasis. Aging Cell. 2013;**12**:635-644. DOI: 10.1111/acel.12085

[63] de Lange T. How shelterin solves the telomere end-protection problem. Cold Spring Harbor Symposia on Quantitative Biology. 2010;**75**:167-177. DOI: 10.1101/sqb.2010.75.017

[64] Matte A, Bertoldi M, Mohandas N, An X, Bugatti A, Brunati AM, Rusnati M, Tibaldi E, Siciliano A, Turrini F, Perrotta S, De Franceschi L. Membrane association of peroxiredoxin-2 in red cells is mediated by the N-terminal cytoplasmic domain of band 3. Free Radical Biology and Medicine. 2013;**55**:27-35. DOI: 10.1016/j.freeradbiomed.2012.10.543

[65] Cho C-S, Yoon HJ, Kim JY, Woo HA, Rhee SG. Circadian rhythm of hyperoxidized peroxiredoxin II is determined by hemoglobin autoxidation and the 20S proteasome in red blood cells. Proceedings of the National Academy of Sciences of the United States of America. 2014;**111**:12043-12048. DOI: 10.1073/pnas.1401100111

[66] Saccoccia F, Di Micco P, Boumis G, Brunori M, Koutris I, Miele AE, Morea V, Sriratana P, Williams DL, Bellelli A, Angelucci F. Moonlighting by different stressors: Crystal structure of the chaperone species of a 2-Cys peroxiredoxin. Structure. 2012;**20**:429-439. DOI: 10.1016/j.str.2012.01.004

[67] Radjainia M, Venugopal H, Desfosses A, Phillips AJ, Yewdall NA, Hampton MB, Gerrard JA, Mitra AK. Cryo-electron microscopy structure of human peroxiredoxin-3 filament reveals the assembly of a putative chaperone. Structure. 2015;**23**:912-920. DOI: 10.1016/j.str.2015.03.019

[68] Nelson KJ, Parsonage D, Karplus PA, Poole LB. Evaluating peroxiredoxin sensitivity toward inactivation by peroxide substrates. Methods in Enzymology. 2013;**527**:21-40. DOI: 10.1016/B978-0-12-405882-8.00002-7

[69] Cox AG, Pearson AG, Pullar JM, Onsson TJJ, Lowther WT, Winterbourn CC, Hampton MB. Mitochondrial peroxiredoxin 3 is more resilient to hyperoxidation than cytoplasmic peroxiredoxins. Movie. 2009;**58**:51-58. DOI: 10.1042/BJ20090242

[70] Woo HA, Kang SW, Kim HK, Yang KS, Chae HZ, Rhee SG. Reversible oxidation of the active site cysteine of peroxiredoxins to cysteine sulfinic acid: Immunoblot detection with antibodies specific for the hyperoxidized cysteine-containing sequence. Journal of Biological Chemistry. 2003;**278**:47361-47364. DOI: 10.1074/jbc.C300428200

Chapter 7

HMGB Proteins from Yeast to Human. Gene Regulation, DNA Repair and Beyond

Vizoso-Vázquez Ángel, Barreiro-Alonso Aida, Rico-Díaz Agustín, Lamas-Maceiras Mónica, Rodríguez-Belmonte Esther, Becerra Manuel, González-Siso María Isabel and Cerdán María Esperanza

Additional information is available at the end of the chapter

http://dx.doi.org/10.5772/intechopen.70126

Abstract

HMGB proteins are characterized for containing one or more HMG-box domains and are well conserved from yeasts to higher eukaryotes. The HMG-box domain is formed by three α-helices with an L-shaped fold. Although HMGB proteins also have cytoplasmic and extracellular functions, they bind to nuclear or mitochondrial DNA in a highly dynamic process that affects chromatin organization. In this review, we mainly focus on HMGB proteins from yeast and their human homologs as functionally involved in DNA repair and transcriptional regulation. Recent research reveals that these proteins participate in epigenetic control of gene expression, aging, disease, or stem-cell biology.

Keywords: nonhistone proteins, epigenetics, transcriptional regulation

1. Introduction

Nucleosomes are fairly stable basic units of DNA packaging. Nevertheless, nucleosomal chromatin is surrounded by a highly dynamic protein pool that allows chromatin remodeling and favors replication, DNA repair, and gene transcription. Among proteins that transiently associate with chromatin are variants of the linker histone H1 family [1–3] and members of the high mobility group (HMG) protein superfamily [4–6]. Although HMG motifs are present in many nuclear proteins, the classification and nomenclature of the considered "canonical"

HMG proteins is organized in three families named HMGA, HMGB, and HMGN, each one having a specific functional domain: the "AT hook" in HMGA, the "HMG-box" in HMGB, and the "nucleosomal binding domain" in HMGN proteins [7].

Some HMGB proteins have been related to nuclear, extranuclear, and extracellular functions during inflammation, cell differentiation, cell migration, and tumor metastasis [8, 9]. Their HMG-box domain contains 65–85 amino acids and has a characteristic L-shaped fold formed by three α-helices with an angle of ≈80° between the two arms. The long arm, or minor wing, is composed by the extended N-terminal strand and third α-helix, while first and second α-helices form the short arm, or major wing (**Figure 1(a)**). Because of protein interaction in the minor groove, DNA-bending and widening of the double helix is produced (**Figure 1(b)**).

There are two broad subfamilies of HMGB-containing proteins, based on structural and phylogenetic studies. One class includes those that bind to distorted DNA with low or without sequence specificity (nonsequence specificity (NSS), HMG-box domains) and have, in general, two or more *in tandem* arranged HMG-box domains [10, 11]. Examples of proteins without sequence specificity are the mammalian Hmgb1-4 and Ubf proteins, Hmgd from *Drosophila*, or Nhp6a and Nhp6b from *Saccharomyces cerevisiae*. Their role is related to chromatin modification, participating in many different functions such as co-activation or silencing of transcription and V(D)J junction recombination. A second class of HMGB-containing proteins binds to DNA by recognizing a specific DNA sequence (sequence specificity (SS), HMG-box domains), and they usually contain a single HMG-box domain [10, 11]. They generally function as transcription factors, only expressed in a few cell types, and they also contain other regulatory associated domain. The determinants for DNA sequence specificity lie mainly in the minor wing of the HMG-box. Examples of this kind of HMGB proteins are the mammalian lymphoid enhancer factor (Lef-1), the sex determining factor (Sry), and the Sry-related HMG-box (SOX) family, or the hypoxic gene repressor (Rox1) from *S. cerevisiae*.

In this review, we focus on HMGB proteins from yeast, as functionally involved in DNA repair and transcriptional regulation, but also in their homologs from multicellular eukaryotes, with special reference to human proteins. Their functions may be modulated by nucleosome positioning and stability [12]. Interestingly, recent findings support that HMGB proteins may also

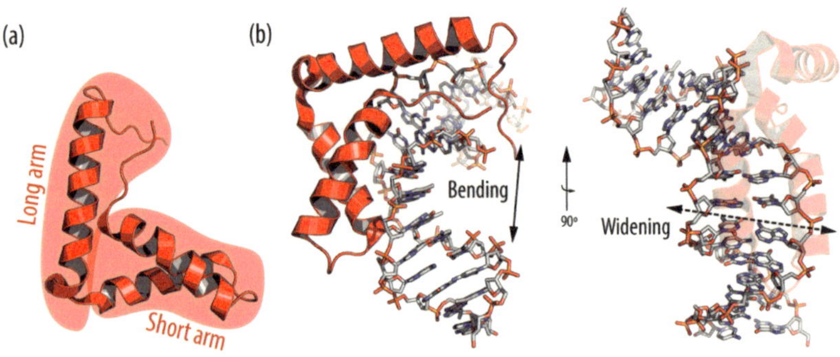

Figure 1. (a) Characteristic HMG-box fold (based on Sox17 protein structure; PDBID: 3F27). (b) Bending and widening produced in the double strand of DNA after protein binding.

http://dx.doi.org/10.5772/intechopen.70126

play diverse roles in epigenetic control, since their interaction with chromatin affects the level of histone modifications [13]. In the light of recently opened research areas, in which HMGB proteins are involved, available knowledge is also discussed.

2. HMGB proteins from *Saccharomyces cerevisiae*

In *S. cerevisiae,* the genes *ABF2, HMO1, IXR1, NHP6A, NHP6B, NHP10,* and *ROX1* encode HMGB proteins [7]. The protein Spp41 also contains a HMG-like motif although homology searches reveal that it is far related to the others. The structural characteristics and functions of these yeast proteins are shown in **Table 1**. Only one HMG-box domain is present in most of them, but Abf2 and Ixr1 have two in tandem "HMG-box" motifs.

With the exception of Rox1 that behaves as a specific transcriptional regulator of the hypoxic yeast regulon [14] and Ixr1 that has a dual function as specific transcription factor and DNA-binding protein without sequence specificity, also participating in DNA repair [15], the other

Protein	Amino acids	Molecular weight (Da)	pI	Aliphatic index	Instability index	Domain position
Abf2	183	21,575	10.24	67.27	42.94	HMG: 42-112 HMG: 115-183 Coil 89-110
Hmo1	246	27,546	9.11	67.35	45.80	HMG: 105-180 PHHR13711: 22-185
Nhp6A	93	10,810	10.40	43.13	39.16	HMG: 20-90 PHHR13711: 7-93
Nhp6B	99	11,485	10.54	37.99	58.30	HMG: 26-96 PTHR13711: 6-99
Nhp10	203	23,858	8.15	68.12	51.57	Coil: 3-24 HMG: 93-159 PTHR13711: 74-182
Rox1	368	41,857	10.46	70.38	62.14	Coil: 90-118 HMG: 9-84
Ixr1	597	67,858	8.36	51.20	70.67	HMG: 360-430 HMG: 433-503 Poly-Q: 3 regions Coil: 292-313 PTHR13711: 1-594

Table 1. Characteristics of HMGB proteins in *Saccharomyces cerevisiae.*

HMGB proteins from *S. cerevisiae* might be considered as chromatin architectural proteins, but with wide influence on gene expression [16]. This is not a HMGB-exclusive mechanism since, in eukaryotes, many other chromatin components, such as histones [17], histone chaperones and modifiers [18], chromatin remodel complexes [19], and long noncoding RNAs [20], affect gene expression by different mechanisms.

Although Abf2 and Ixr1 are considered paralogs, resulting from the whole genome duplication in an ancestor of *Saccharomyces*, the function of Abf2 is not related to transcriptional regulation of hypoxic regulons. Abf2 is a mitochondrial DNA-binding protein involved in mitochondrial DNA replication and recombination [21, 22]. *In vivo*, PKA-mediated phosphorylation of Abf2 during glucose repression may regulate its functions on maintaining mitochondrial DNA content during the shift from gluconeogenic to fermentative growth [21].

Hmo1 is not considered a specific transcriptional factor either, although it regulates rDNA transcription from RNA polymerase I promoters and also regulates start site selection of ribosomal protein genes by RNA polymerase II [23–25].

Nhp10 (alias Hmo2) is a nonessential subunit of the INO80 chromatin remodeling complex, and it affects telomere maintenance via recombination [26, 27].

Nhp6a and Nhp6b are also paralogs and functionally redundant [28], they bind to and remodel nucleosomes [29, 30], and both are required for transcriptional initiation fidelity of some tRNA genes [31]. Their protein levels increase in response to DNA replication stress [32]. Besides, Nhp6a and Nhp6b acting on chromatin tightly repress histone expression; paradoxically, histone gene overexpression in the double *nhp6aΔ nhp6bΔ* mutant is compensated by downregulation of translation, finally determining a histone-decreased phenotype to avoid the toxic effect of histone overproduction [33].

Although few data are available about Ssp41 functions, it has been associated with chromatin remodeling [34], transcription, and RNA processing [35, 36]. Besides, overexpression causes chromosomal instability [36] and under hypoxia, it is rapidly exported to the cytosol [34].

An intriguing question is whether the *S. cerevisiae* HMGB proteins contribute altogether to regulate specific cell functions. An interesting perspective comes from the terms "environmental stress response" (ESR) or "common environmental response" (CER). These terms refer to adaptive yeast responses against acute changes in diverse environmental parameters (e.g., O_2, osmolarity, nutrients, pH, UV, etc.), which evoke a common transcriptional response, initially devoted to mitigate the deleterious effect of the specific stressor, but principally to balance cell energetics and to coordinate progression through the cell cycle [37]. We have summarized the information available in SGD about protein interactants of HMGB proteins from *S. cerevisiae* (http://www.yeastgenome.org/ as accessed date February 22, 2017) and used this information to construct a interactome network using STRING facilities (http://string-db.org/). **Figure 2** shows that this network statistically has significantly more interactions among HMG-box proteins and their previously reported partners than randomly expected, with a *p*-value < 0.01 according to STRING analysis. This result suggests that yeast HMGB proteins are related, not only structurally but also as a functional group. **Table 2** summarizes GOTerm enrichment analysis among the components of this network and their statistical significances evaluated by false discovery rate (FDR) according to STRING analysis [120].

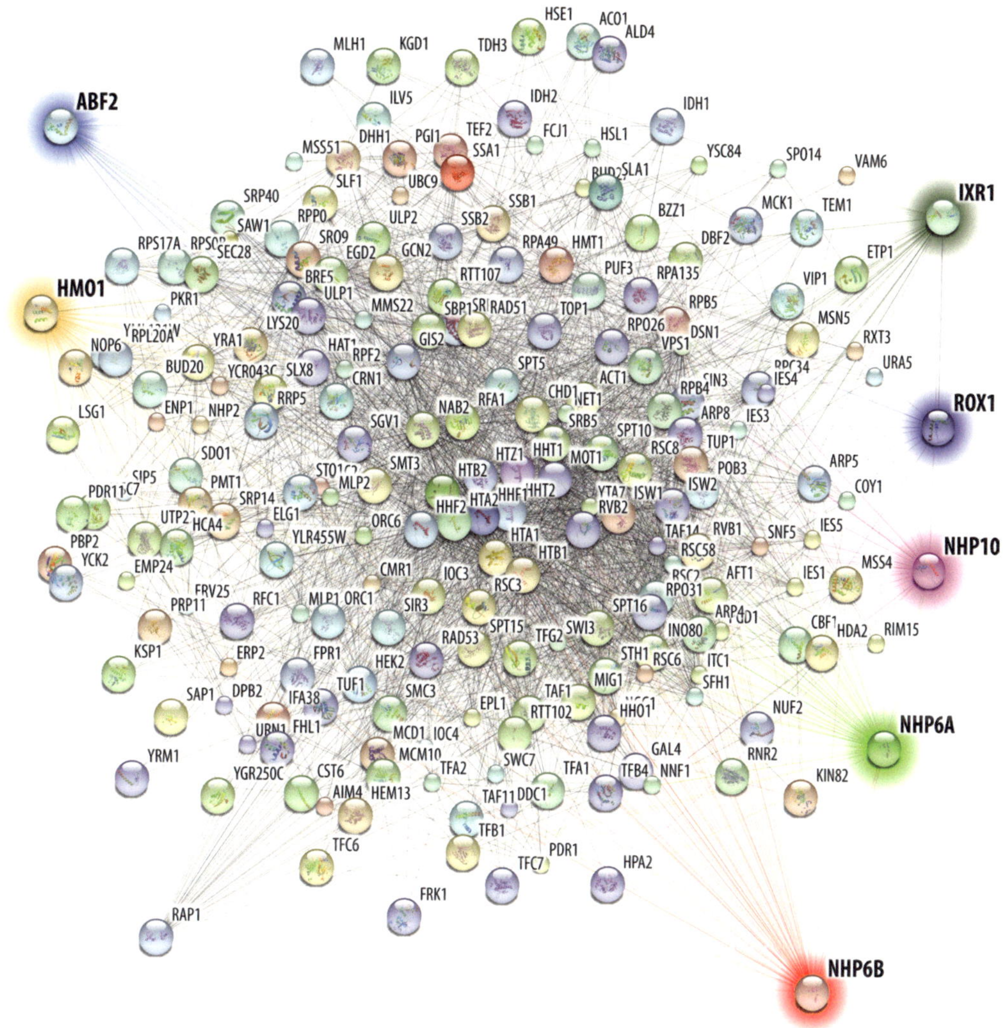

Figure 2. Network of yeast HMGB interactants according to STRING analysis.

References to the existence of interplay between the response to hypoxia, oxidative stress, and mitochondrial function have been reported, i.e., it is known that when cells experience hypoxia, up- or downregulation of an important number of oxygen-regulated genes in yeast depends on an active mitochondrial respiratory chain [38]. Treatment with antimycin A (respiration inhibitor) or oxygen deprivation cause downregulation of networks involved in the G1/S transition of the cell cycle as well as of those involved in energetically costly programs of ribosomal biogenesis and protein synthesis [37]. Similar regulation occurs in the response to DNA stress [39–41], and therefore, a wide gene-regulatory response might engage the functions of the HMGB proteins coordinately. **Figure 3** summarizes the participation of HMGB proteins from *S. cerevisiae* in functional responses against external (nutrient availability, oxidants, oxygen levels, DNA damaging agents) or internal (replicative stress) stressors.

Pathway ID	Biological function; pathway description	Observed gene count	False discovery rate
GO.0006325	Chromatin organization	50	4.50E-27
GO.0010468	Regulation of gene expression	81	5.85E-25
GO.0051171	Regulation of nitrogen compound metabolic process	84	5.85E-25
GO.0051276	Chromosome organization	63	5.85E-25
GO.0006355	Regulation of transcription, DNA-templated	71	3.30E-24
GO.0051252	Regulation of RNA metabolic process	72	3.30E-24
GO.0071824	Protein-DNA complex subunit organization	39	1.27E-22
GO.0043933	Macromolecular complex subunit organization	78	1.24E-21
GO.0090304	Nucleic acid metabolic process	95	1.53E-21
GO.0034728	Nucleosome organization	27	3.55E-21
GO.0006338	Chromatin remodeling	26	3.00E-20
GO.0006351	Transcription, DNA-templated	61	3.69E-19
GO.0006974	Cellular response to DNA damage stimulus	43	5.34E-19
GO.0006333	Chromatin assembly or disassembly	24	7.13E-19
GO.0006281	DNA repair	39	2.21E-18
GO.0016568	Chromatin modification	36	3.02E-18
GO.0006259	DNA metabolic process	47	1.11E-17
GO.0010467	Gene expression	82	1.39E-15
GO.0016070	RNA metabolic process	78	1.73E-15
GO.0006357	Regulation of transcription from RNA polymerase II promoter	45	8.94E-14
GO.0006323	DNA packaging	16	1.73E-11
GO.0006366	Transcription from RNA polymerase II promoter	29	1.99E-11
GO.0043044	ATP-dependent chromatin remodeling	14	5.72E-11
GO.0006950	Response to stress	54	6.10E-10
GO.0016458	Gene silencing	22	1.12E-09

Pathway ID	Biological function; pathway description	Observed gene count	False discovery rate
GO.0006354	DNA-templated transcription, elongation	16	1.21E-09
GO.0040029	Regulation of gene expression, epigenetic	23	1.75E-09
GO.0050896	Response to stimulus	67	3.76E-09
GO.0006342	Chromatin silencing	21	4.01E-09
GO.0071103	DNA conformation change	16	9.65E-08
GO.0016584	Nucleosome positioning	7	1.67E-07
GO.0007049	Cell cycle	49	2.67E-07
GO.0018193	Peptidyl-amino acid modification	18	5.44E-07
GO.0022607	Cellular component assembly	47	7.03E-07
GO.0065004	Protein-DNA complex assembly	17	9.19E-07
GO.1902589	Single-organism organelle organization	58	9.77E-07
GO.0042766	Nucleosome mobilization	7	9.81E-07
GO.0018205	Peptidyl-lysine modification	15	1.06E-06
GO.0022402	Cell cycle process	43	2.30E-06
GO.0006337	Nucleosome disassembly	8	2.35E-06
GO.0031498	Chromatin disassembly	8	2.35E-06
GO.0006368	Transcription elongation from RNA polymerase II promoter	12	2.40E-06
GO.0006302	Double-strand break repair	16	2.65E-06
GO.0098781	ncRNA transcription	12	4.52E-06
GO.0000122	Negative regulation of transcription from RNA polymerase II	19	7.75E-06
GO.0006383	Transcription from RNA polymerase III promoter	9	9.65E-06
GO.0000723	Telomere maintenance	12	4.18E-05
GO.0006360	Transcription from RNA polymerase I promoter	9	6.50E-05
GO.0009303	rRNA transcription	8	0.000151
GO.0016570	Histone modification	13	0.000222

Table 2. GOTerm enrichment in the interactome network depicted in **Figure 2**.

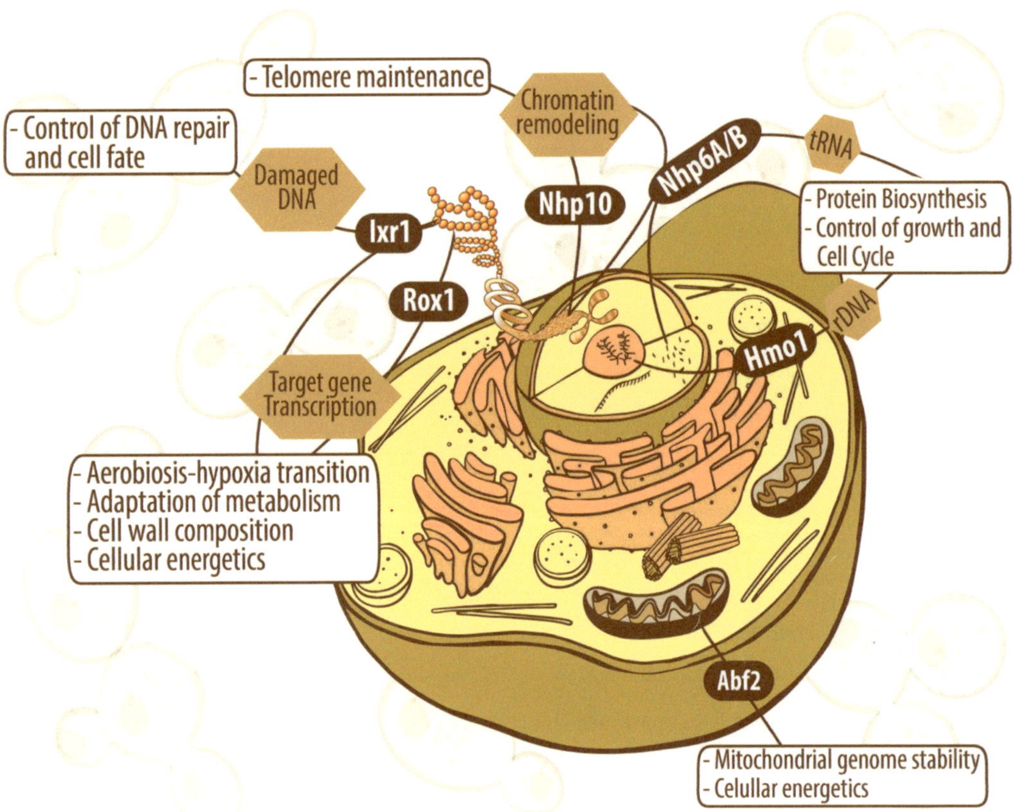

Figure 3. Orchestrated action of *S. cerevisiae* HMGB proteins in cellular responses to stress.

3. HMGB proteins from other yeasts

Although the complete sequences of a huge number of genomes from yeast and fungi are available, functional studies of HMGB proteins are not very frequent and only a few HMGB homologs have been characterized so far.

In *Yarrowia lipolytica,* YlMhb1, the homologous of Abf2 from *S. cerevisiae,* compacts mitochondrial DNA *in vitro*. Phenotypic analysis of a *mhb1Δ* strain reveals a large decrease in the mitochondrial DNA copy number and also shows that the protein protects the mitochondrial genome against mutagenic events. Like Abf2, YlMhb1 has two HMG-box domains [42]. In *Candida parapsilosis,* the homologous of Abf2 has been named Gcf1 and diverse experimental data support its role in the maintenance of the *C. parapsilosis* mitochondrial genome; in contrast to Abf2 and YlMhb1, Gcf1contains a coiled-coil domain and a single high-mobility HMG-box domain [43]. A similar structure is observed in *Candida albicans* [44].

In *C. albicans*, proteins with DNA-binding activity and high similarity to Nhp6 promote changes in chromatin structure, which are involved in hypha-specific gene regulation [45].

Regarding the Rox1 homolog in *Kluyveromyces lactis*, its molecular function, synteny, and HMG-box structural features were shown to be different from that of *S. cerevisiae* [46, 47]. The *KlROX1* gene from *K. lactis* does not regulate the hypoxic response in this yeast neither interacts with the components of the general corepressor factor (Tup1-Ssn6) that mediates the transcriptional repression exerted by Rox1 in *S. cerevisiae*. However, KlRox1 mediates the response to metals [47].

Although a low number of functional data is available, we may speculate that in yeasts the functions of "architectural" HMGB proteins are probably more conserved than those with functions as specific transcriptional factors. This is also predictable considering that transcriptional factors are among the proteins more strongly diverged between yeasts [48].

4. HMGB proteins in multicellular organisms

In multicellular eukaryotes, a large number of proteins contain HMG boxes, most of which are transcription factors that contain a single HMG-box [49], although some may have up to 6 HMG-box domains, like Ubf1 [50]. According to the classification from Bustin [7], "canonical" chromatin HMGB proteins represent a subgroup that invariably contains two in tandem HMG boxes. A model for the phylogenesis of HMGB genes in metazoan suggests that these two HMG boxes have their origin in the duplication of an ancient single HMG-box; even those which are part of HMG-box transcription factors might evolve from this ancestral ProtoBox [51].

Transcription factors (including SOX factors) are the most divergent group of HMG-box proteins in humans, whereas in plants the chromosomal HMGB-type proteins are most variable [52]. In plants, HMG-box proteins classify into four groups: HMGB-type proteins, structure-specific recognition protein 1 (SSRP1), proteins containing 3 HMG-box domains (3xHMG-box), and proteins that contain both an AT-rich interaction domain (ARID) and an HMG-box domain (ARID/HMG). These latter two groups are apparently specific for plants [52]. Conversely, HMG-box containing transcription factors such as Sry, a sex-determining factor that is necessary for testes development [53], Lef-1, which regulates gene expression during cell differentiation [54], and the SOX family are presumably not present in plants [52].

Table 3 resumes the homologies found between *S. cerevisiae* and human HMGB groups using the YeastMine facility "Yeast gene-human homolog(s)-Disease" (http://yeastmine.yeastgenome.org/yeastmine/begin.do accessed on date February 25, 2017) and completed with functional data from SGD (http://www.yeastgenome.org/) and associated human diseases. **Figure 4** summarizes the structural and phylogenetic relationships between several HMGB proteins from *S. cerevisiae* and their human homologs.

Yeast	*H. sapiens*	Associated human diseases
Rox1	Sox1	
	Sox10	Peripheral demyelinating neuropathy, central dysmyelination, Waardenburg syndrome, and Hirschsprung disease
	Sox11	Mental retardation, autosomal dominant 27
	Sox12	
	Sox13	
	Sox14	
	Sox15	
	Sox17	Vesicoureteral reflux
	Sox18	Hypotrichosis-lymphedema-telangiectasia-renal defect syndrome
	Sox2	Microphthalmia, syndromic 3
	Sox21	
	Sox3	Mental retardation, X-linked
	Sox30	
	Sox4	
	Sox5	
	Sox6	
	Sox7	
	Sox8	
	Sox9	Campomelic dysplasia
	Sry	46,Xx sex reversal 1
Ixr1	Hmg20a	
	Hmg20b	
	Smarce1	Susceptibility to familial meningioma
	Sp110	Susceptibility to *Mycobacterium tuberculosis*
	Sp140	
	Tfam	
	Ubtf	
	Ubtfl1	
Abf2	Tfam	
Hmo1	Hmg20a	
	Hmg20b	
	Smarce1	Susceptibility to familial meningioma
	Sp110	Susceptibility to *Mycobacterium tuberculosis*
	Sp140	
	Tfam	
	Ubtf	
	Ubtfl1	
Nhp6a/b	Hmg20a	
	Hmg20b	
	Hmgb1	
	Hmgb3	Microphthalmia, syndromic 13
	Smarce1	Susceptibility to familial meningioma
	Sp110	Susceptibility to *Mycobacterium tuberculosis*
	Sp140	
	Tfam	
	Ubtf	
	Ubtfl1	

Table 3. Human homologs to HMGB yeast proteins and associated diseases.

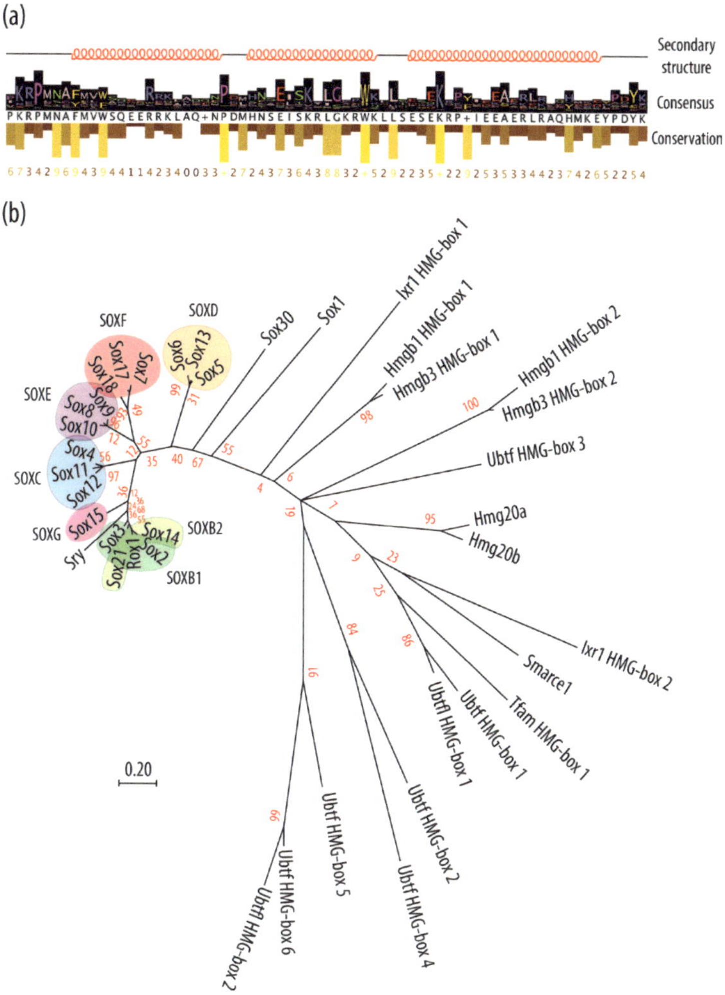

Figure 4. Molecular phylogenetic analysis of HMG-box domains by maximum likelihood method. (a) Characteristic HMG-box conservation. The evolutionary history was inferred by using the maximum likelihood method based on the JTT matrix-based model [118]. (b) The tree with the highest log likelihood (−3421.5683) is shown. Initial tree(s) for the heuristic search were obtained automatically by applying Neighbor-Join and BioNJ algorithms to a matrix of pairwise distances estimated using a JTT model, and then selecting the topology with superior log likelihood value. The tree is drawn to scale, with branch lengths measured in the number of substitutions per site. The analysis involved 39 amino acid sequences. All positions containing gaps and missing data were eliminated. There were a total of 63 positions in the final dataset. Evolutionary analyses were conducted in MEGA7 [119].

5. Mechanisms of transcriptional regulation mediated by HMGB proteins

5.1. Direct binding to target promoters

In *S. cerevisiae*, Rox1 is a DNA-binding protein with an HMG-box domain that binds to the consensus sequence YYYATTGTTCTC present in the promoter regions of genes related to hypoxia, causing a DNA bending of 90° in the double strand [55, 56]. Up to one-third of the *S. cerevisiae* hypoxic genes are transcriptionally repressed during aerobic growth by Rox1, through the recruitment of the general corepressor complex Ssn6/Tup1 [14, 57]. In several promoters, this repression is synergic with the caused by regulator Mot3 [58]. *ROX1* expression is dependent of oxygen and heme levels in the cell, since its transcription is under the control of Hap1 [59], and therefore, it is induced aerobically [60]. In addition to aerobic upregulation produced by Hap1, the *ROX1* expression is counterbalanced by self-repression, to avoid cell toxic effects produced by an eventual overexpression. At low oxygen levels, the Rox1 protein levels rapidly decay by degradation, since it is labile in these conditions, and because the *ROX1* gene is no longer transcribed. Under normoxic (aerobic) conditions, the heme-activated Hap1 complex increases *ROX1* expression, allowing in turn Rox1 repression of hypoxic genes. In hypoxia, the situation is reversed, since the low levels of Rox1 allow derepression. The genes that are under the control of Rox1, either directly by the protein binding to their promoter regions, or indirectly through signal transduction pathways, are those related to efficient metabolism under low oxygen levels, ergosterol and heme synthesis, cell wall maintenance, or electron chain transport [61]. The genes repressed simultaneously by Mot3 and Rox1 preferentially encode proteins of the cell wall and plasma membrane; cell conjugation-related genes are negatively regulated by both factors and by osmotic stress [62]. During anaerobiosis, the histone deacetylase and global repressor complex Rpd3 act at the promoter of the anaerobic gene *DAN1* to antagonize the chromatin-mediated repression caused by Mot3 and Rox1 and chromatin remodeling by Swi/Snf is necessary for expression [63].

The first report about the participation of Ixr1 in the yeast hypoxic response was the aerobic repression of the *COX5B* gene, which encodes the hypoxic isoform of the subunit Vb of the mitochondrial complex cytochrome c oxidase [64]. Ixr1 also regulates other hypoxic genes like *TIR1*, a cell wall mannoprotein of the serine-alanine-rich protein family [65] and *HEM13*, which encodes the enzyme coproporphyrinogen III oxidase in the heme biosynthetic pathway [66]. The whole set of genes that are regulated by Ixr1 during the hypoxic response was determined in a genome-wide approach [67]. Hypoxic genes are also regulated by oxidative stress. Indeed, reactive oxygen species (ROS) induce expression of *CYC7* and *COX5B* through an Ixr1-independent mechanism that diminishes the access of Rox1 to its promoter targets [68].

A cross-regulation between Rox1 and Ixr1 in the yeast hypoxic response has been reported [66]. In aerobiosis, low levels of *IXR1* expression are maintained by Rox1 repression and during hypoxia Ixr1 auto-enhances *IXR1* expression [66]. Ixr1 is also required for hypoxic repression of *ROX1*. Binding to specific regions of the *ROX1, IXR1, HEM13*, and *TIR1* promoters were probed *in vitro* and *in vivo* [66, 69]. Ixr1 is also known by binding to cisplatin-DNA adducts with high affinity [70]. We have recently evidenced that functional specialization of the 2 HMG boxes, which are present in Ixr1, may explain its dual function. Regulation of transcription and DNA repair is achieved through differential recognition of specific regulatory sequences in the target promoters, or DNA disturbances caused by cisplatin treatment [15].

Rox1 from *S. cerevisiae* is homologous to the SOX family of transcriptional factors from human (**Table 2**) and other metazoan, from which SRY was the founding member. In vertebrates, there are more than 20 SOX genes characterized, which originate through a process of duplication and divergence [71], and they play important roles in tissue homeostasis, organogenesis, and cell fate decision during developmental processes (thoroughly reviewed by Ref. [72]).

http://dx.doi.org/10.5772/intechopen.70126

For most mammals, SRY is the only member of the SOXA group [73]. SOXB1 group (SOX1, SOX2, and SOX3) participates in neural, lens, and ear development; SOXB2 group (SOX14 and SOX21) in neuronal differentiation SOXC group (SOX4, SOX11, and SOX12) in nervous system development and retinal differentiation; and SOXD group (SOX5, SOX6, and SOX13) in chondrocyte differentiation, cartilage formation, and neural development. SOXE group (SOX8, SOX9, and SOX10) is involved in primary sex determination and neural development, and SOXF group (SOX7, SOX17, and SOX18) in cardiac, vascular, and lymphatic development [72]. The SOXG group has only one member in mammals, and SOX15 involved in skeletal muscle regeneration [72, 74]. Besides, SOX4 and SOX11 are involved in tumorigenesis, and SOX7, SOX17, and SOX18 in endoderm development [72]. **Figure 5** summarizes the functions of these human SOX factors.

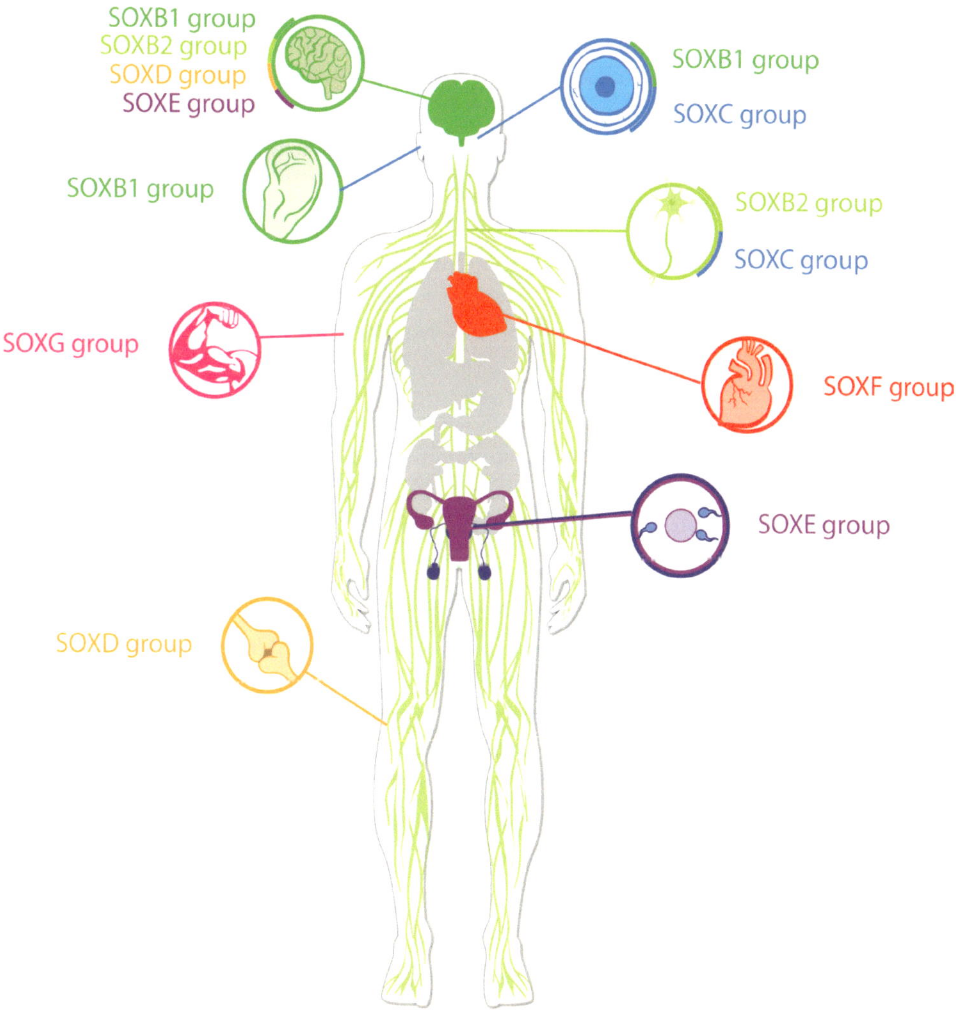

Figure 5. Functional groups of human SOX factors.

SOX proteins are highly dynamic regulators of cell functions due to their nucleocytoplasmic shuttling properties [75]. However, because of their low affinity for DNA binding, and despite SOX proteins usually have their own C-terminal activation/repression domain, they are committed to recruit partner proteins to fulfill their transcriptional regulatory task [76]. Homo- and heterodimerization of SOX proteins is also a mechanism used for the formation of these regulatory complexes [77]. SOX proteins also interact with signaling effectors, Wnt/β-catenin being one of the most studied signaling pathways [78]. Different molecular complexes of SOX factors and their partner proteins are formed along developmental processes. Besides, these specific interactions are usually dependent on posttranslational modifications of SOX proteins like phosphorylation, acetylation, SUMOylation, and ubiquitination [72].

5.2. Other mechanisms for transcriptional regulation

The HMGB proteins that are not classified as transcriptional factors also influence transcription by different mechanisms, which affect chromatin. Since these HMGB proteins are very dynamic in their interactions and have no DNA sequence specificity, they usually help transcription factors or cofactors to bind to their cognate sites by bending the DNA molecule, but are rarely retained within the formed complexes [79].

In plants, HMGB proteins contribute to transcriptional regulation by functional interaction with certain transcription factors like Dof2 [80]. In mammals, Hmgb1 alters the structure and stability of the canonical nucleosome in a nonenzymatic, ATP-independent way to facilitate strong binding of estrogen receptor to their regulatory elements [81].

HMGB proteins also interact with nucleosomes to promote their sliding or other chromatin remodeling processes [79]. Yeast Nhp6a, Nhp6b, and Hmo1 proteins stimulate the sliding activity of the yeast remodeler complex SWI/SNF, while octamer transfer and transient exposure of nucleosomal DNA catalyzed by this complex are only stimulated by Hmo1. Hmo1 also favors the sliding activity of the ISW1a complex [82].

Hmo1 in yeasts and the upstream binding factor (Ubf) in mammals function as cofactors in RNA polymerase I transcription and therefore are essential for transcription of the rRNA genes *in vivo*, but also have more generalized roles in chromatin structure. Binding of Ubf to human rRNA genes is accompanied by a reduction in core histone binding at the same sequences [83, 84], and a similar mechanism has been described for its ortholog Hmo1 in yeast [25]. Similarly, mammalian cells lacking Hmgb1 and yeast *nhp6* mutants contain a reduced amount of core, linker, and variant histones [85]. Consequently, the reduced number of nucleosomes produces a global increment of transcription and affects the relative expression of about 10% of genes [85].

Finally, HMGB proteins have been involved in the selection of modified histone variants. Studies carried out in mouse showed that conditional inactivation of Ubf is also accompanied by recruitment of H3K9me3, which reveals its function in the epigenetic control of gene expression [86].

6. Mechanisms of DNA repair mediated by HMGB proteins

The three HMG families (A, B, N) are involved in the four major DNA repair pathways. HMGB proteins contribute to nucleotide excision repair (NER), base excision repair (BER), double-strand break repair (DSBR), and mismatch repair (MMR), but with specific particularities (reviewed in Ref. [87]). The first report about participation of HMGB proteins in DNA repair was the identification of Hmgb1 binding to the major DNA lesions formed in cells treated with cisplatin, which are repaired by the NER pathway [88]. In general, the effects of HMGB proteins on DNA repair are achieved by different mechanisms. First, they contribute to modulate chromatin compaction and nucleosome occupancy; through interactions with chromatin-modifying enzymes and energy-dependent remodeling complexes, HMGB proteins favor or avoid the access of the repair machinery to altered DNA. Second, HMGB proteins can also regulate repair by direct modulation of the enzymatic activities and/or mechanistic steps implied in the diverse repair pathways. Third, acting as transcriptional regulators, HMGB proteins may change the expression levels of genes involved in DNA repair processes.

Hmgb1 and many other HMGB proteins (e.g., Ubf, Lef-1, Sry, and human mtTFA) inhibit NER [87]. If Hmgb1 binds first to a cisplatin adduct, the replication protein A (hRPA), necessary for NER repair, cannot displace it, thus potentially inhibiting repair [89]. On the contrary, Hmgb1 stimulates *in vitro* NER of triplex DNA interstrand crosslinks, caused by psoralen, by facilitating the interaction with components of this pathway [83, 90].

Hmgb1 coimmunoprecipitates with proteins from the BER pathway, including Ape1, Fen-1, and Pol-beta, and *in vitro*, modulates the deoxyribose phosphate lyase activity of Pol-beta [91].

Also *in vitro*, purified Hmgb1 binds to the ends of the double-strand breaks, similarly to the Ku proteins, and stimulates kinase and ligase activities required for DBSR of these lesions [92, 93]. Oppositely, in yeast, the HMGB protein Hmo1 must be evicted, along with core histones, for efficient DSBR [94].

Hmgb1 and Hmgb2 form part of a pentameric "damage-sensing" complex (also including heat shock protein 70, protein disulfide-isomerase Erp60, and glyceraldehyde3-phosphate dehydrogenase) specifically recruited to nonnatural nucleosides *in vivo* as part of the MMR pathway [95]. *In vitro*, Hmgb1 also interacts with the MMR proteins Msh2 and Mlh1 and cooperates with the replication protein A to mediate the exonuclease I activity that creates a gap, which is filled in by DNA polymerase, and finally, the broken strands are sealed by DNA ligase [96]. In yeast, following Nhp6a interaction to DNA, the mismatch repair complex Msh2-Msh6 is excluded from binding, unless a mismatch is present. *In vitro* the complex Msh2-Msh6-Nhp6a is stable and responsive to ATP on mismatched substrates [97].

Other important connection between Hmgb1 and DNA repair comes from the observation that this protein interacts with p53 *in vitro* and *in vivo*, stimulating p53 binding to sequence-specific recognition sites as well as to cisplatin-modified DNA [98, 99]. p53 directly impacts the activity of various DNA-repair systems, and besides, it halts cell cycle, thus allowing the repair machineries to restore genome stability [100].

7. HMGB proteins at the forefront of cutting-edge research

Recent publications on HMGB proteins reveal that these proteins are becoming a focus of interest due to their participation in cellular processes of great importance for humankind like epigenetic control of gene expression, aging, disease, or regenerative cellular therapies.

An interesting research field concerning HMGB proteins is their function replacing histones under specific conditions. In eukaryotic chromatin, histone H1 associates with the linker DNA in the nucleosome core particle to stabilize the higher-order chromatin structure and to modulate the ability of specific regulatory factors to access their final targets. It has been demonstrated that in *S. cerevisiae* Hmo1 might replace histone H1 and protect linker DNA from nuclease digestion, creating a less dynamic chromatin environment that depends on its lysine-rich domain. This lysine-rich extension is unusual in other HMGB proteins, which have an acidic domain instead [101, 102].

Environmental changes, sensed through signaling cascades, regulate chromatin organization, thus contributing to gene expression and, ultimately, cell adaptation to external stimulus. These responses are related to cell fate and aging. In yeast, the nutrient-dependent target-of-rapamycin complex 1 (TORC1) pathway and histone H3 collaborate to retain HMGB proteins within the nucleus, and in this way, they increase longevity [103].

The role of HMGB proteins remodeling chromatin on a genome-wide scale relates to the onset of several human diseases. Two chromatin structural proteins, CCCTC-binding factor (Ctcf) and high mobility group protein B2 (Hmgb2), regulate pathologic transcription in myocytes during heart disease [104]. The response of macrophages to inflammation starts by nucleosome loss and cell lacking Hmgb1 contains 20% less nucleosomes and has a specific transcription pattern. In a mouse model, unstimulated Hmgb1-/- macrophages activate transcriptional pathways associated with cell migration and chemotaxis. Wild-type macrophages, under lipopolysaccharide (LPS)/interferon (IFN)-γ exposure, rapidly secrete Hmgb1 and reduce their histone content [105].

Hmgb1 is overexpressed in many types of cancer, including those of etiology based on oxidative damage [8], and frequently, Hmgb1 expression increases with tumor stage and metastasis. In the pediatric acute lymphoblastic leukemia, autophagy is regarded as a mechanism that underlies chemoresistance. Since autophagy depends on the Hmgb1 translocation from nucleus to cytoplasm, this protein is a good target of study in order to overcome the problem [106]. It has been found that Hmgb1 expression is inversely correlated with semaphorin 3A expression, a suppressor of angiogenesis and cell migration. The epigenetic mechanism causing semaphorin 3A repression by Hmgb1 implies that it promotes heterochromatin formation and decreased occupancy of acetylated histones at the semaphorin 3A locus [107].

Other remarkable function of HMGB proteins, yet not fully understood, is their participation in telomere maintenance, studied in yeast [108] plants [109] and notoriously in animals [110], because of their implications in cancer development. The telomerase that conserves telomere structures is formed by a catalytic protein subunit (telomerase reverse transcriptase (TERT)) and an RNA subunit (telomerase RNA, TR), and both physically interact with Hmgb1 *in vitro*. Knockout of the HMGB1 gene in mouse embryonic fibroblasts (MEFs) causes chromosomal abnormalities, enhanced localization of γ-H2AX at telomeres, moderate shortening of telomere lengths, and lower telomerase activity compared to the wild-type cells. Oppositely, knockout of the HMGB2 gene elevates telomerase activity, which reveals the intricate interplay of these proteins in chromosome stability and cancer [110].

Evidences linking HMGB proteins with stem cell biology and cellular reprograming are also found. Sox factors participate in embryonic pluripotent cell differentiation; Oct4 interacts with Sox2 to maintain pluripotency or with Sox17 to promote endoderm commitment [111]. Expression of Hmgb2 changes notably at different time points during embryogenesis [112] and controls the differentiation of neural stem cells into neurons, astrocytes, and oligodendrocytes. Besides, several Sox factors [113, 114] and also chromatin HMGB proteins [115] are involved in back-reprograming differentiated cells into stem cells. Hmgb1 was also proposed as an efficient stem cell recruiter with tissue-regenerating roles; it was able to induce stem cell transmigration through an endothelial barrier or to capture in muscle the stem cells injected into the general circulation [116]. In murine and human mammary cancer stem cells, Hmgb1 promotes self-renewal of these cells [117], which are responsible for tumor progression, metastases, resistance to therapy, and tumor recurrence. Therefore, HMGB proteins are clues in the search of more effective cancer therapies and cellular regenerative treatments.

Acknowledgements

Funding is acknowledged both from the Instituto de Salud Carlos III under Grant Agreement no. PI14/01031 cofinanced by FEDER and from Xunta de Galicia (Consolidación D.O.G. X-12-2016. Contract Number: 2016/012). Aida Barreiro-Alonso was funded by a predoctoral fellowship from Plan I2C Xunta de Galicia-2013 (Spain). Agustín Rico-Diaz was funded by a predoctoral fellowship from Plan I2C Xunta de Galicia-2012 (Spain). We thank STRING facilities and development.

Author details

Vizoso-Vázquez Ángel, Barreiro-Alonso Aida, Rico-Díaz Agustín,
Lamas-Maceiras Mónica, Rodríguez-Belmonte Esther, Becerra Manuel,
González-Siso María Isabel and Cerdán María Esperanza*

*Address all correspondence to: esper.cerdan@udc.es

Facultade de Ciencias, Centro de Investigacións Científicas Avanzadas (CICA) e Instituto de Investigacións Biomédicas Coruña (INIBIC), Universidade da Coruña, Grupo EXPRELA, A Coruña, Spain

References

[1] Kasinsky HE, Lewis JD, Dacks JB, Ausio J. Origin of H1 linker histones. FASEB Journal: Official Publication of the Federation of American Societies for Experimental Biology. 2001;**15**:34-42. DOI: 10.1096/fj.00-0237rev

[2] Lever MA, Th'ng JP, Sun X, Hendzel MJ. Rapid exchange of histone H1.1 on chromatin in living human cells. Nature. 2000;**408**:873-876. DOI: 10.1038/35048603

[3] Misteli T, Gunjan A, Hock R, Bustin M, Brown DT. Dynamic binding of histone H1 to chromatin in living cells. Nature. 2000;**408**:877-881. DOI: 10.1038/35048610

[4] Bustin M. Regulation of DNA-dependent activities by the functional motifs of the high-mobility-group chromosomal proteins. Molecular and Cellular Biology. 1999;**19**:5237-5246

[5] Bianchi ME, Agresti A. HMG proteins: Dynamic players in gene regulation and differentiation. Current Opinion in Genetics and Development. 2005;**15**:496-506. DOI: S0959-437X (05)00137-1

[6] Gerlitz G, Hock R, Ueda T, Bustin M. The dynamics of HMG protein-chromatin interactions in living cells. Biochemistry and Cell Biology = Biochimie Et Biologie Cellulaire. 2009;**87**:127-137. DOI: 10.1139/O08-110

[7] Bustin M. Revised nomenclature for high mobility group (HMG) chromosomal proteins. Trends in Biochemical Sciences. 2001;**26**:152-153. DOI: S0968-0004(00)01777-1

[8] Barreiro-Alonso A, Lamas-Maceiras M, Rodriguez-Belmonte E, Vizoso-Vazquez A, Quindos M, Cerdan ME. High mobility group b proteins, their partners, and other redox sensors in ovarian and prostate cancer. Oxidative Medicine and Cellular Longevity. 2015;**2016**:5845061. DOI: 10.1155/2016/5845061

[9] Hock R, Furusawa T, Ueda T, Bustin M. HMG chromosomal proteins in development and disease. Trends in Cell Biology. 2007;**17**:72-79. DOI: S0962-8924(06)00338-2

[10] Bustin M, Reeves R. High-mobility-group chromosomal proteins: Architectural components that facilitate chromatin function. Progress in Nucleic Acid Research and Molecular Biology. 1996;**54**:35-100

[11] Grosschedl R, Giese K, Pagel J. HMG domain proteins: Architectural elements in the assembly of nucleoprotein structures. Trends in Genetics. 1994;**10**:94-100

[12] Bonaldi T, Langst G, Strohner R, Becker PB, Bianchi ME. The DNA chaperone HMGB1 facilitates ACF/CHRAC-dependent nucleosome sliding. The EMBO Journal. 2002;**21**:6865-6873

[13] Lange SS, Mitchell DL, Vasquez KM. High mobility group protein B1 enhances DNA repair and chromatin modification after DNA damage. Proceedings of the National Academy of Sciences of the United States of America. 2008;**105**:10320-10325. DOI: 10.1073/pnas.0803181105

[14] Kastaniotis AJ, Zitomer RS. Rox1 mediated repression. Oxygen dependent repression in yeast. Advances in Experimental Medicine and Biology. 2000;**475**:185-195

[15] Vizoso-Vázquez A, Lamas-Maceiras M, Fernández-Leiro R, Rico-Díaz A, Becerra M, Cerdán ME. Dual function of Ixr1 in transcriptional regulation and recognition of cisplatin-DNA adducts is caused by differential binding through its two HMG-boxes. Biochimica Et Biophysica Acta. 2017;**1860**:256-269. DOI: S1874-9399(16)30254-1

[16] Misteli T. Protein dynamics: Implications for nuclear architecture and gene expression. Science (New York, N.Y.). 2001;**291**:843-847

[17] Wu J, Grunstein M. 25 Years after the nucleosome model: Chromatin modifications. Trends in Biochemical Sciences. 2000;**25**:619-623. DOI: S0968-0004(00)01718-7

[18] Hondele M, Ladurner AG. The chaperone-histone partnership: For the greater good of histone traffic and chromatin plasticity. Current Opinion in Structural Biology. 2011;**21**:698-708. DOI: 10.1016/j.sbi.2011.10.003

[19] Swygert SG, Peterson CL. Chromatin dynamics: Interplay between remodeling enzymes and histone modifications. Biochimica Et Biophysica Acta. 2014;**1839**:728-736. DOI: 10.1016/j.bbagrm.2014.02.013

[20] Fatica A, Bozzoni I. Long non-coding RNAs: New players in cell differentiation and development. Nature Reviews. Genetics. 2014;**15**:7-21. DOI: 10.1038/nrg3606

[21] Cho JH, Lee YK, Chae CB. The modulation of the biological activities of mitochondrial histone Abf2p by yeast PKA and its possible role in the regulation of mitochondrial DNA content during glucose repression. Biochimica Et Biophysica Acta - Gene Structure and Expression. 2001;**1522**:175-186. DOI: 10.1016/S0167-4781(01)00333-5

[22] Macalpine DM, Perlman PS, Butow RA. The high mobility group protein Abf2p influences the level of yeast mitochondrial DNA recombination intermediates *in vivo*. Proceedings of the National Academy of Sciences of the United States of America. 1998;**95**:6739-6743. DOI: 10.1073/pnas.95.12.6739

[23] Gadal O, Labarre S, Boschiero C, Thuriaux P. Hmo1, an HMG-box protein, belongs to the yeast ribosomal DNA transcription system. The EMBO Journal. 2002;**21**:5498-5507. DOI: 10.1093/emboj/cdf539

[24] Kasahara K, Ki S, Aoyama K, Takahashi H, Kokubo T. *Saccharomyces cerevisiae* HMO1 interacts with TFIID and participates in start site selection by RNA polymerase II. Nucleic Acids Research. 2008;**36**:1343-1357. DOI: 10.1093/nar/gkm1068

[25] Merz K, Hondele M, Goetze H, Gmelch K, Stoeckl U, Griesenbeck J. Actively transcribed rRNA genes in S. cerevisiae are organized in a specialized chromatin associated with the high-mobility group protein Hmo1 and are largely devoid of histone molecules. Genes and Development. 2008;**22**:1190-1204. DOI: 10.1101/gad.466908

[26] Conaway RC, Conaway JW. The INO80 chromatin remodeling complex in transcription, replication and repair. Trends in Biochemical Sciences. 2009;**34**:71-77. DOI: 10.1016/j.tibs.2008.10.010

[27] Kasahara K, Ohtsuki K, Ki S, Aoyama K, Takahashi H, Kobayashi T, Shirahige K, Kokubo T. Assembly of regulatory factors on rRNA and ribosomal protein genes in *Saccharomyces cerevisiae*. Molecular and Cellular Biology. 2007;**27**:6686-6705. DOI: MCB.00876-07

[28] Byrne KP, Wolfe KH. The yeast gene order browser: Combining curated homology and syntenic context reveals gene fate in polyploid species. Genome Research. 2005;**15**:1456-1461. DOI: gr.3672305

[29] Paull TT, Johnson RC. DNA looping by Saccharomyces cerevisiae high mobility group proteins NHP6A/B. Consequences for nucleoprotein complex assembly and chromatin condensation. The Journal of Biological Chemistry. 1995;**270**:8744-8754

[30] Stillman DJ. Nhp6: A small but powerful effector of chromatin structure in *Saccharomyces cerevisiae*. Biochimica Et Biophysica Acta. 2010;**1799**:175-180. DOI: 10.1016/j.bbagrm.2009.11.010

[31] Braglia P, Dugas SL, Donze D, Dieci G. Requirement of Nhp6 proteins for transcription of a subset of tRNA genes and heterochromatin barrier function in *Saccharomyces cerevisiae*. Molecular and Cellular Biology. 2007;**27**:1545-1557. DOI: MCB.00773-06

[32] Tkach JM, Yimit A, Lee AY, Riffle M, Costanzo M, Jaschob D, et al. Dissecting DNA damage response pathways by analysing protein localization and abundance changes during DNA replication stress. Nature Cell Biology. 2012;**14**:966-976. DOI: 10.1038/ncb2549

[33] Durano D, Lukacs A, Di Felice F, Micheli G, Camilloni G. A novel role for Nhp6 proteins in histone gene regulation in *Saccharomyces cerevisiae*. The International Journal of Biochemistry and Cell Biology. 2017;**83**:76-83. DOI: S1357-2725(16)30399-5

[34] Dastidar RG, Hooda J, Shah A, Cao TM, Henke RM, Zhang L. The nuclear localization of SWI/SNF proteins is subjected to oxygen regulation. Cell and Bioscience. 2012;**2**:30. DOI: 10.1186/2045-3701-2-30

[35] Maddock JR, Weidenhammer EM, Adams CC, Lunz RL, Woolford JL Jr. Extragenic suppressors of *Saccharomyces cerevisiae prp4* mutations identify a negative regulator of PRP genes. Genetics. 1994;**136**:833-847

[36] Zhu J, Heinecke D, Mulla WA, Bradford WD, Rubinstein B, Box A, Haug JS ,Li R. Single-cell based quantitative assay of chromosome transmission fidelity. G3 (Bethesda, Md.). 2015;**5**:1043-1056. DOI: 10.1534/g3.115.017913

[37] Lai LC, Kissinger MT, Burke PV, Kwast KE. Comparison of the transcriptomic "stress response" evoked by antimycin A and oxygen deprivation in *Saccharomyces cerevisiae*. BMC Genomics. 2008;**9**:627. DOI: 10.1186/1471-2164-9-627

[38] Poyton RO, Ball KA, Castello PR. Mitochondrial generation of free radicals and hypoxic signaling. Trends in Endocrinology and Metabolism: TEM. 2009;**20**:332-340. DOI: 10.1016/j.tem.2009.04.001

[39] Causton HC, Ren B, Koh SS, Harbison CT, Kanin E, Jennings EG, Lee TI, True HL, Lander ES, Young RA. Remodeling of yeast genome expression in response to environmental changes. Molecular Biology of the Cell. 2001;**12**:323-337

[40] Gasch AP, Spellman PT, Kao CM, Carmel-Harel O, Eisen MB, Storz G, Botstein D, Brown PO. Genomic expression programs in the response of yeast cells to environmental changes. Molecular Biology of the Cell. 2000;**11**:4241-4257

[41] Mazumder A, Pesudo LQ, McRee S, Bathe M, Samson LD. Genome-wide single-cell-level screen for protein abundance and localization changes in response to DNA damage in *S. cerevisiae*. Nucleic Acids Research. 2013;**41**:9310-9324. DOI: 10.1093/nar/gkt715

[42] Bakkaiova J, Marini V, Willcox S, Nosek J, Griffith JD, Krejci L, Tomaska L. Yeast mitochondrial HMG proteins: DNA-binding properties of the most evolutionarily divergent component of mitochondrial nucleoids. Bioscience Reports. 2016;**36**. DOI: 10.1042/BSR20150275

[43] Miyakawa I, Okamuro A, Kinsky S, Visacka K, Tomaska L, Nosek J. Mitochondrial nucleoids from the yeast Candida parapsilosis: Expansion of the repertoire of proteins associated with mitochondrial DNA. Microbiology. 2009;**155**:1558-1568. DOI: 10.1099/mic.0.027474-0

[44] Visacka K, Gerhold JM, Petrovicova J, Kinsky S, Jöers P, Nosek J, Sedman J, Tomaska L. Novel subfamily of mitochondrial HMG box-containing proteins: Functional analysis of Gcf1p from *Candida albicans*. Microbiology. 2009;**155**:1226-1240. DOI: 10.1099/mic.0.025759-0

[45] Kim S, Wolyniak MJ, Staab JF, Sundstrom P. A 368-base-pair cis-acting HWP1 promoter region, HCR, of *Candida albicans* confers hypha-specific gene regulation and binds architectural transcription factors Nhp6 and Gcf1p. Eukaryotic Cell. 2007;**6**:693-709. DOI: EC.00341-06

[46] Fang ZA, Wang GH, Chen AL, Li YF, Liu JP, Li YY, Bolotin-Fukuhara M, Bao WG. Gene responses to oxygen availability in *Kluyveromyces lactis*: An insight on the evolution of the oxygen-responding system in yeast. PLoS One. 2009;**4**:e7561. DOI: 10.1371/journal.pone.0007561

[47] Rodríguez-Torres AM, Maceiras ML, Belmonte ER, Naveira LN, Calvo MB, Cerdan ME. KlRox1p contributes to yeast resistance to metals and is necessary for KlYCF1 expression in the presence of cadmium. Gene. 2012;**497**:27-37. DOI: 10.1016/j.gene.2012.01.050

[48] Bussereau F, Casaregola S, Lafay JF, Bolotin-Fukuhara M. The *Kluyveromyces lactis* repertoire of transcriptional regulators. FEMS Yeast Research. 2006;**6**:325-335. DOI: 10.1111/j.1567-1364.2006.00028.x

[49] Thomas JO, Travers AA. HMG1 and 2, and related 'architectural' DNA-binding proteins. Trends in Biochemical Sciences. 2001;**26**:167-174. DOI: S0968-0004(01)01801-1

[50] Russell J, Zomerdijk JC. RNA-polymerase-I-directed rDNA transcription, life and works. Trends in Biochemical Sciences. 2005;**30**:87-96. DOI: S0968-0004(04)00322-6

[51] Sessa L, Bianchi ME. The evolution of High Mobility Group Box (HMGB) chromatin proteins in multicellular animals. Gene. 2007;**387**:133-140. DOI: S0378-1119(06)00585-3

[52] Stros M, Launholt D, Grasser KD. The HMG-box: A versatile protein domain occurring in a wide variety of DNA-binding proteins. Cellular and Molecular Life Sciences: CMLS. 2007;**64**:2590-2606. DOI: 10.1007/s00018-007-7162-3

[53] Berta P, Hawkins JR, Sinclair AH, Taylor A, Griffiths BL, Goodfellow PN, Fellous M. Genetic evidence equating SRY and the testis-determining factor. Nature. 1990;**348**:448-450. DOI: 10.1038/348448A0

[54] Hurlstone A, Clevers H. T-cell factors: Turn-ons and turn-offs. The EMBO Journal. 2002;**21**:2303-2311. DOI: 10.1093/emboj/21.10.2303

[55] Deckert J, Khalaf RA, Hwang S, Zitomer RS. Characterization of the DNA binding and bending HMG domain of the yeast hypoxic repressor Rox1. Nucleic Acids Research. 1999;**27**:3518-3526. DOI: 10.1093/nar/27.17.3518

[56] Xin H, Taudte S, Kallenbach NR, Limbach MP, Zitomer RS. DNA binding by single HMG box model proteins. Nucleic Acids Research. 2000;**28**:4044-4050

[57] Mennella TA, Klinkenberg LG, Zitomer RS. Recruitment of Tup1-Ssn6 by yeast hypoxic genes and chromatin-independent exclusion of TATA binding protein. Eukaryotic Cell. 2003;**2**:1288-1303

[58] Klinkenberg LG, Mennella TA, Luetkenhaus K, Zitomer RS. Combinatorial repression of the hypoxic genes of *Saccharomyces cerevisiae* by DNA binding proteins Rox1 and Mot3. Eukaryotic Cell. 2005;**4**:649-660. DOI: 4/4/649; 10.1128/EC.4.4.649-660.2005

[59] Keng T. *HAP1* and *ROX1* form a regulatory pathway in the repression of *HEM13* transcription in *Saccharomyces cerevisiae*. Molecular and Cellular Biology. 1992;**12**:2616-2623

[60] Deckert J, Torres AM, Hwang SM, Kastaniotis AJ, Zitomer RS. The anatomy of a hypoxic operator in *Saccharomyces cerevisiae*. Genetics. 1998;**150**:1429-1441

[61] Ter Linde JJ, Steensma HY. A microarray-assisted screen for potential Hap1 and Rox1 target genes in Saccharomyces cerevisiae. Yeast (Chichester, England). 2002;**19**:825-840. DOI: 10.1002/yea.879

[62] Martinez-Montanes F, Rienzo A, Poveda-Huertes D, Pascual-Ahuir A, Proft M. Activator and repressor functions of the Mot3 transcription factor in the osmostress response of *Saccharomyces cerevisiae*. Eukaryotic Cell. 2013;**12**:636-647. DOI: 10.1128/EC.00037-13

[63] Sertil O, Vemula A, Salmon SL, Morse RH, Lowry CV. Direct role for the Rpd3 complex in transcriptional induction of the anaerobic *DAN/TIR* genes in yeast. Molecular and Cellular Biology. 2007;**27**:2037-2047. DOI: MCB.02297-06

[64] Lambert JR, Bilanchone VW, Cumsky MG. The ORD1 gene encodes a transcription factor involved in oxygen regulation and is identical to IXR1, a gene that confers cisplatin sensitivity to *Saccharomyces cerevisiae*. Proceedings of the National Academy of Sciences. 1994;**91**:7345-7349

[65] Bourdineaud JP, De Sampaio G, Lauquin GJ. A Rox1-independent hypoxic pathway in yeast. Antagonistic action of the repressor Ord1 and activator Yap1 for hypoxic expression of the *SRP1/TIR1* gene. Molecular Microbiology. 2000;**38**:879-890. DOI: mmi2188

[66] Castro-Prego R, Lamas Maceiras M, Soengas P, Carneiro I, González-Siso I, Cerdán ME. Regulatory factors controlling transcription of *Saccharomyces cerevisiae IXR1* by oxygen levels: A model of transcriptional adaptation from aerobiosis to hypoxia implicating *ROX1*and *IXR1*cross-regulation. The Biochemical Journal. 2010;**425**:235-243. DOI: 10.1002/yea.975

[67] Vizoso-Vázquez A, Lamas-Maceiras M, Becerra M, González-Siso MI, Rodríguez-Belmonte E, Cerdán ME. Ixr1p and the control of the *Saccharomyces cerevisiae* hypoxic response. Applied Microbiology and Biotechnology. 2012;**94**:173-184. DOI: 10.1007/s00253-011-3785-2

[68] Liu J, Barrientos A. Transcriptional regulation of yeast oxidative phosphorylation hypoxic genes by oxidative stress. Antioxidants and Redox Signaling. 2013;**19**:1916-1927. DOI: 10.1089/ars.2012.4589

[69] Castro-Prego R, Lamas-Maceiras M, Soengas P, Fernández-Leiro R, Carneiro I, Becerra M, González-Siso MI, Cerdán ME. Ixr1p regulates oxygen-dependent *HEM13* transcription. FEMS Yeast Research. 2010;**10**:309-321. DOI: 10.1111/j.1567-1364.2010.00616.x

[70] Brown SJ, Kellett PJ, Lippard SJ. Ixr1, a yeast protein that binds to platinated DNA and confers sensitivity to cisplatin. Science. 1993;**261**:603-605

[71] Bowles J, Schepers G, Koopman P. Phylogeny of the SOX family of developmental transcription factors based on sequence and structural indicators. Developmental Biology. 2000;**227**:239-255. DOI: 10.1006/dbio.2000.9883

[72] She ZY, Yang WX. SOX family transcription factors involved in diverse cellular events during development. European Journal of Cell Biology. 2015;**94**:547-563. DOI: 10.1016/j.ejcb.2015.08.002

[73] Kashimada K, Koopman P. Sry: The master switch in mammalian sex determination. Development. 2010;**137**:3921-3930. DOI: 10.1242/dev.048983

[74] Ito M. Function and molecular evolution of mammalian Sox15, a singleton in the SoxG group of transcription factors. The International Journal of Biochemistry and Cell Biology. 2010;**42**:449-452. DOI: 10.1016/j.biocel.2009.10.023

[75] Malki S, Boizet-Bonhoure B, Poulat F. Shuttling of SOX proteins. The International Journal of Biochemistry and Cell Biology. 2010;**42**:411-416. DOI: 10.1016/j.biocel.2009.09.020

[76] Kondoh H, Kamachi Y. SOX-partner code for cell specification: Regulatory target selection and underlying molecular mechanisms. The International Journal of Biochemistry and Cell Biology. 2010;**42**:391-399. DOI: 10.1016/j.biocel.2009.09.003

[77] Lefebvre V. The SoxD transcription factors—Sox5, Sox6, and Sox13—are key cell fate modulators. The International Journal of Biochemistry and Cell Biology. 2010;**42**:429-432. DOI: 10.1016/j.biocel.2009.07.016

[78] Bernard P, Harley VR. Acquisition of SOX transcription factor specificity through protein-protein interaction, modulation of Wnt signaling and post-translational modification. The International Journal of Biochemistry and Cell Biology. 2010;**42**:400-410. DOI: 10.1016/j.biocel.2009.10.017

[79] Agresti A, Bianchi ME. HMGB proteins and gene expression. Current Opinion in Genetics and Development. 2003;**13**:170-178. DOI: S0959437X03000236

[80] Grasser M, Christensen JM, Peterhansel C, Grasser KD. Basic and acidic regions flanking the HMG-box domain of maize HMGB1 and HMGB5 modulate the stimulatory effect on the DNA binding of transcription factor Dof2. Biochemistry. 2007;**46**:6375-6382. DOI: 10.1021/bi6024947

[81] Joshi SR, Sarpong YC, Peterson RC, Scovell WM. Nucleosome dynamics: HMGB1 relaxes canonical nucleosome structure to facilitate estrogen receptor binding. Nucleic Acids Research. 2012;**40**:10161-10171. DOI: 10.1093/nar/gks815

[82] Hepp MI, Alarcon V, Dutta A, Workman JL, Gutierrez JL. Nucleosome remodeling by the SWI/SNF complex is enhanced by yeast high mobility group box (HMGB) proteins. Biochimica Et Biophysica Acta. 2014;**1839**:764-772. DOI: 10.1016/j.bbagrm.2014.06.014

[83] Lange SS, Reddy MC, Vasquez KM. Human HMGB1 directly facilitates interactions between nucleotide excision repair proteins on triplex-directed psoralen interstrand crosslinks. DNA Repair. 2009;**8**:865-872. DOI: 10.1016/j.dnarep.2009.04.001

[84] Gagnon-Kugler T, Langlois F, Stefanovsky V, Lessard F, Moss T. Loss of human ribosomal gene CpG methylation enhances cryptic RNA polymerase II transcription and disrupts ribosomal RNA processing. Molecular Cell. 2009;**35**:414-425. DOI: 10.1016/j.molcel.2009.07.008

[85] Celona B, Weiner A, Di Felice F, Mancuso FM, Cesarini E, Rossi RL, et al. Substantial histone reduction modulates genome wide nucleosomal occupancy and global transcriptional output. PLoS Biology. 2011;**9**:e1001086. DOI: 10.1371/journal.pbio.1001086

[86] Hamdane N, Stefanovsky VY, Tremblay MG, Nemeth A, Paquet E, Lessard F, Sanij E, Hannan R, Moss T. Conditional inactivation of upstream binding factor reveals its epigenetic functions and the existence of a somatic nucleolar precursor body. PLoS Genetics. 2014;**10**:e1004505. DOI: 10.1371/journal.pgen.1004505

[87] Reeves R. High mobility group (HMG) proteins: Modulators of chromatin structure and DNA repair in mammalian cells. DNA Repair. 2015;**36**:122-136. DOI:10.1016/j.dnarep.2015.09.015

[88] Zamble DB, Mu D, Reardon JT, Sancar A, Lippard SJ. Repair of cisplatin—DNA adducts by the mammalian excision nuclease. Biochemistry. 1996;**35**:10004-10013. DOI: 10.1021/bi960453+

[89] Patrick SM, Turchi JJ. Human replication protein A preferentially binds cisplatin-damaged duplex DNA *in vitro*. Biochemistry. 1998;**37**:8808-8815. DOI: 10.1021/bi9730590

[90] Lange SS, Vasquez KM. HMGB1: The jack-of-all-trades protein is a master DNA repair mechanic. Molecular Carcinogenesis. 2009;**48**:571-580. DOI: 10.1002/mc.20544

[91] Liu Y, Prasad R, Wilson SH. HMGB1: Roles in base excision repair and related function. Biochimica Et Biophysica Acta. 2010;**1799**:119-130. DOI: 10.1016/j.bbagrm.2009.11.008

[92] Yumoto Y, Shirakawa H, Yoshida M, Suwa A, Watanabe F, Teraoka H. High mobility group proteins 1 and 2 can function as DNA-binding regulatory components for DNA-dependent protein kinase *in vitro*. Journal of Biochemistry. 1998;**124**:519-527

[93] Stros M, Cherny D, Jovin TM. HMG1 protein stimulates DNA end joining by promoting association of DNA molecules via their ends. European Journal of Biochemistry. 2000;**267**:4088-4097. DOI: ejb1450

[94] Panday A, Xiao L, Grove A. Yeast high mobility group protein HMO1 stabilizes chromatin and is evicted during repair of DNA double strand breaks. Nucleic Acids Research. 2015;**43**:5759-5770. DOI: 10.1093/nar/gkv498

[95] Krynetski EY, Krynetskaia NF, Bianchi ME, Evans WE. A nuclear protein complex containing high mobility group proteins B1 and B2, heat shock cognate protein 70, ERp60, and glyceraldehyde-3-phosphate dehydrogenase is involved in the cytotoxic response to DNA modified by incorporation of anticancer nucleoside analogues. Cancer Research. 2003;**63**:100-106

[96] Zhang Y, Yuan F, Presnell SR, Tian K, Gao Y, Tomkinson AE, Gu L, Li GM. Reconstitution of 5′-directed human mismatch repair in a purified system. Cell. 2005;**122**:693-705. DOI: S0092-8674(05)00646-X

[97] Labazi M, Jaafar L, Flores-Rozas H. Modulation of the DNA-binding activity of *Saccharomyces cerevisiae* MSH2-MSH6 complex by the high-mobility group protein NHP6A, *in vitro*. Nucleic Acids Research. 2009;**37**:7581-7589. DOI: 10.1093/nar/gkp649

[98] Imamura T, Izumi H, Nagatani G, Ise T, Nomoto M, Iwamoto Y, Kohno K. Interaction with p53 enhances binding of cisplatin-modified DNA by high mobility group 1 protein. The Journal of Biological Chemistry. 2001;**276**:7534-7540. DOI: 10.1074/jbc.M008143200

[99] Jayaraman L, Moorthy NC, Murthy KG, Manley JL, Bustin M, Prives C. High mobility group protein-1 (HMG-1) is a unique activator of p53. Genes and Development. 1998;**12**:462-472

[100] Williams AB, Schumacher B. p53 in the DNA-damage-repair process. Cold Spring Harbor Perspectives in Medicine. 2016;**6**. DOI: 10.1101/cshperspect.a026070

[101] Panday A, Grove A. Yeast HMO1: Linker histone reinvented. Microbiology and Molecular Biology Reviews: MMBR. 2016;**81**:e00037-16. DOI: e00037-16

[102] Panday A, Grove A. The high mobility group protein HMO1 functions as a linker histone in yeast. Epigenetics and Chromatin. 2016;**9**:13. DOI: 10.1186/s13072-016-0062-8

[103] Chen H, Workman JJ, Strahl BD, Laribee RN. Histone H3 and TORC1 prevent organelle dysfunction and cell death by promoting nuclear retention of HMGB proteins. Epigenetics and Chromatin. 2016;**9**:34. DOI: 10.1186/s13072-016-0083-3

[104] Monte E, Rosa-Garrido M, Karbassi E, Chen H, Lopez R, Rau CD, et al. Reciprocal regulation of the cardiac epigenome by chromatin structural proteins Hmgb and Ctcf: Implications for transcriptional regulation. The Journal of Biological Chemistry. 2016;**291**:15428-15446. DOI: 10.1074/jbc.M116.719633

[105] De Toma I, Rossetti G, Zambrano S, Bianchi ME, Agresti A. Nucleosome loss facilitates the chemotactic response of macrophages. Journal of Internal Medicine. 2014;**276**:454-469. DOI: 10.1111/joim.12286

[106] Kong Q, Xu LH, Xu W, Fang JP, Xu HG. HMGB1 translocation is involved in the transformation of autophagy complexes and promotes chemoresistance in leukaemia. International Journal of Oncology. 2015;**47**:161-170. DOI: 10.3892/ijo.2015.2985

[107] Nehil M, Paquette J, Tokuyasu T, McCormick F. High mobility group box 1 promotes tumor cell migration through epigenetic silencing of semaphorin 3A. Oncogene. 2014;**33**:5151-5162. DOI: 10.1038/onc.2013.459

[108] Hu Y, Tang HB, Liu NN, Tong XJ, Dang W, Duan YM, et al. Telomerase-null survivor screening identifies novel telomere recombination regulators. PLoS Genetics. 2013;**9**:e1003208. DOI: 10.1371/journal.pgen.1003208

[109] Schrumpfova PP, Fojtova M, Mokros P, Grasser KD, Fajkus J. Role of HMGB proteins in chromatin dynamics and telomere maintenance in *Arabidopsis thaliana*. Current Protein and Peptide Science. 2011;**12**:105-111. DOI: CPPS-100

[110] Polanska E, Dobsakova Z, Dvorackova M, Fajkus J, Stros M. HMGB1 gene knockout in mouse embryonic fibroblasts results in reduced telomerase activity and telomere dysfunction. Chromosoma. 2012;**121**:419-431. DOI: 10.1007/s00412-012-0373-x

[111] Merino F, Ng CK, Veerapandian V, Scholer HR, Jauch R, Cojocaru V. Structural basis for the SOX-dependent genomic redistribution of OCT4 in stem cell differentiation. Structure. 2014;**22**:1274-1286. DOI: 10.1016/j.str.2014.06.014

[112] Abraham AB, Bronstein R, Chen EI, Koller A, Ronfani L, Maletic-Savatic M, Tsirka SE. Members of the high mobility group B protein family are dynamically expressed in embryonic neural stem cells. Proteome Science. 2013;**11**:18. DOI: 10.1186/1477-5956-11-18

[113] Aksoy I, Jauch R, Eras V, Chng WB, Chen J, Divakar U, Ng CK, Kolatkar PR, Stanton LW. Sox transcription factors require selective interactions with Oct4 and specific transactivation functions to mediate reprogramming. Stem Cells. 2013;**31**:2632-2646. DOI: 10.1002/stem.1522

[114] Sarkar A, Hochedlinger K. The sox family of transcription factors: Versatile regulators of stem and progenitor cell fate. Cell Stem Cell. 2013;**12**:15-30. DOI: 10.1016/j.stem.2012.12.007

[115] Li M, Sun L, Luo Y, Xie C, Pang Y, Li Y. High-mobility group box 1 released from astrocytes promotes the proliferation of cultured neural stem/progenitor cells. International Journal of Molecular Medicine. 2014;**34**:705-714. DOI: 10.3892/ijmm.2014.1820

[116] Palumbo R, Bianchi ME. High mobility group box 1 protein, a cue for stem cell recruitment. Biochemical Pharmacology. 2004;**68**:1165-1170. DOI: 10.1016/j.bcp.2004.03.048

[117] Conti L, Lanzardo S, Arigoni M, Antonazzo R, Radaelli E, Cantarella D, Calogero RA, Cavallo F. The noninflammatory role of high mobility group box 1/Toll-like receptor 2 axis in the self-renewal of mammary cancer stem cells. FASEB Journal: Official Publication of the Federation of American Societies for Experimental Biology. 2013;**27**:4731-4744. DOI: 10.1096/fj.13-230201

[118] Jones DT, Taylor WR, Thornton JM. The rapid generation of mutation data matrices from protein sequences. Computer Applications in the Biosciences: CABIOS. 1992;**8**:275-282

[119] Kumar S, Stecher G, Tamura K. MEGA7: Molecular Evolutionary Genetics Analysis version 7.0 for bigger datasets. Molecular Biology and Evolution. 2016;**33**:1870-1874. DOI: 10.1093/molbev/msw054

[120] Szklarczyk D, Morris JH, Cook H, Kuhn M, Wyder S, Simonovic M, Santos A, Doncheva NT, Roth A, Bork P, Jensen LJ, von Mering C. The STRING database in 2017: Quality-controlled protein-protein association networks, made broadly accessible. Nucleic Acids Research. 2017;**45**:D362–D368. DOI: 10.1093/nar/gkw937

Section 3

Yeast Versus Plant - Unknown Territory

Chapter 8

Endophytic Yeast and Hosts: A Mutualistic Association Friendly to the Environment

Esperanza del Pilar Infante Luna

Additional information is available at the end of the chapter

http://dx.doi.org/10.5772/intechopen.70326

Abstract

Recent studies have shown that endophytic yeasts benefit their host, which has stimulated their use in different applications in agribusiness. The research has focused on evaluating the effectiveness of handling these yeasts to solve problems such as biocontrol of pathogens, plant growth and/or improvements in the quality of fruits and vegetables. However, in order to obtain information that contributes to the selection and the implementation of a yeast able to interact with a broader spectrum of hosts and to help solve postharvest problems, it is necessary to deepen the knowledge on the association of these symbionts and to establish possible changes in the host, the issues that are covered in this chapter. The results show that the endophytic yeasts can generate structural changes in the host as a starting point for further applied research and to propose other mechanisms of action.

Keywords: biocontrol, endophytic yeast, mode of action, mutualism, postharvest

1. Introduction

The relationship between plants and microorganisms has been classified as a symbiosis; however, when referring to endophytic yeasts, this association takes a mutualistic character. While the plant is providing the yeast a propitious space to live, the yeast offers benefits to the plant, which are mainly related to the biological control of pathogens, encouraging their use as an alternative method for the management of postharvest diseases of fruits and vegetables [1].

Nevertheless, the knowledge regarding the dynamics of host colonization by the endophyte to understand this mutualistic relationship as well as the evaluation of the inoculated host is still limited. Isaeva et al. [2] state that the research on endophytic yeasts has not been carried out in a systematic way, so the existing information is incomplete. They also identify the need to know the distributional patterns and biological properties of endophytic yeast, in order to understand the ecological characteristics of these yeasts and propose solutions to various postharvest problems.

The fact that endophyte yeasts can live in the host involves studying the dynamics of colonization within the host and establishing whether it is affected by providing a habitat for the yeast surviving, so it is necessary to use alternative methodologies that allow visualizing both the yeast and the host, as well as changes inside it.

Accordingly, the results obtained by implementing techniques of microscopy and magnetic resonance imaging (MRI) in order to evaluate the interaction between a host and an endophyte yeast are explained below. These pieces of evidence allow to deepen the knowledge of this mutualistic relationship and to propose another mode of action of the yeasts in which these indirectly contribute to prolonging the useful life of the host.

2. Endophytic yeasts and plants: a mutualistic action

Etymologically, the word endophyte means "within the plant." This definition encompasses a wide variety of residents and hosts, this last including bacteria, fungi, insects and algae among others [3]. Among the definitions proposed for the term endophyte is "Fungus that colonizes plant tissue without causing any immediate negative effect" [4]. Even so, some authors consider that this definition excludes other microorganisms such as bacteria and algae. In this context, Stone et al. [5] argue that a more wide-ranging definition should emphasize the asymptomatic nature of the infection without taking into account a particular group of organisms. That is why Petrini [6] explains endophyte from a topographical perspective: "An endophyte colonizes and can live inside the living tissues of it is host without causing damage."

Xin et al. [7] ponder all these aspects and characterize endophytic yeast as: "Unicellular fungi that reproduce asexually by budding—without a hyphal phase or with a reduced hyphal phase—and can live in their host without generating apparent harm." Pieces of research show that these yeasts can be isolated from different parts of plants (see **Table 1**).

Yeast	Isolated from	References
Williopsis saturnus	Maize (*Zea mays* L.) roots	Nassar et al. [10]
Wild poplar strain 1 (WP1)	Wild cottonwood (*Populus trichocarpa*)	Xin et al. [7]
PTD 2	Stems of hybrid poplar (*Populus trichocarpa* x *Populus deltoides*)	Xin et al. [7]
Candida guilliermondii	Heterograft tomato crop (HGTC)	Celis et al. [24]

Table 1. Some endophytic yeast reported.

In recent years, there has been an increase in research on how endophytic yeast benefits the host; it has been established that in some cases, it contributes to the protection against pathogens. Therefore, it is possible to use them successfully as agents for biological control [8, 9]. Also, some studies have shown that these yeasts foster the growth of plants by means of bringing out auxins, as reported by Nassar et al. [10], who isolated the endophytic yeast *Williopsis saturnus* and found that it is capable of producing indole-3-acetic acid (IAA), a growth hormone. In addition, Zhao et al. [11] discovered that exogenous administration of *Pichia guilliermondii* improved the postharvest lifetime and the quality of cherry tomato fruits stored.

This association between plant and microorganisms is denominated symbiosis, a term coined by Anton De Bary as: "The association, at least for part of its life cycle, between two or more specifically different organisms" [12]. For the host plant, this relationship can be positive (mutualism); neutral (neutralism), or negative (parasitism or competition). For the symbiotic microorganism, the association can be positive (mutualism, commensalism, or parasitism), neutral (neutralism), or negative when there is competition. A symbiosis is successful provided that it involves at least the following three events: (i) the symbiont's entrance into the tissues; (ii) their colonization and (iii) the expression of one of the symbiotic relationships mentioned above. The symbiont must be able to have a relationship with the host to establishing a compatible interaction, which implies that it overcomes or manipulates the host defense system [13].

It has been verified that in the case of endophytic yeast, the association is closer to mutualism than to parasitism [2] since yeast can bring to the plant several of the above benefits mentioned. On the other hand, yeasts as copiotroph organisms find in the host plant the nutrients and the suitable environment for their development. Here the question is: how do you experimentally identify whether an endophytic yeast is related in a mutualistic way to its host? It could be answered if we adapt Sieber's proposal [14] of using the Koch's four postulates, modified as follows:

1. The appearance of an endophyte should be associated with a benefit to the host.
2. The endophyte should be isolated from the tissue in which the benefit was observed and grown in a culture medium.
3. The endophyte that has grown in the culture medium should generate the same benefit when it is reintroduced into a host free of the endophyte.
4. Then, again the endophyte should be isolated from the experimentally inoculated host.

In order to identify new endophytic yeast, it is possible to apply these postulates experimentally.

Concerning asymptomatic colonization, characteristic of endophytic yeasts, Schulz et al. [3] suggested a hypothesis in which the absence of negative symptoms is associated with a balance of antagonists: host and endophyte. The endophytes have mechanisms to infect and colonize the host; this, in turn, responds with its defense system. The balance between the "infection system" and the "defense system" generates an asymptomatic interaction;

if the balance is broken, diseases can occur for the host or death of the symbiont. However, the verification of this balance, which is an experimental challenge, is not solved in the study of endophytes yet.

3. Endophytic yeasts and their projection in agro-industry

During the postharvest period, the quality of fruits and vegetables is deteriorated due to different factors: manipulation and improper storage, metabolic events, and phytopathogen attacks generating economic losses of more than 25% of the total production in industrialized countries and more than 50% in developing countries [15, 16].

In the case of fruits, most of these losses are caused by the attack of several fungal pathogens, controlled mainly with synthetic fungicides, which has generated concern regarding possible health risks derived from the consumption of food treated with agrochemicals [17], as a consequence, the demand of organic fruits and vegetables has increased. To deal with this need, healthier and environmentally friendly strategies have been evaluated to control the attack of plant pathogens and to maintain the quality of fruits and vegetables, in that context, microbial antagonists, such as yeasts have emerged as a viable option [18].

To understand how the yeast can be used to solve this problem, we can identify different interactions with the host and with the phytopathogen. In relation to the host, the yeast can colonize the fruit surface for long periods; some of them produce extracellular polysaccharides that contribute to the fruit survival and to restrict the growth of pathogens; they can use nutrients from the environment and proliferate at a high rate. In addition, their activity does not involve the production of toxic metabolites and are less affected by fungicides [1, 19]. When a yeast colonizes the internal tissues of the host without generating damage or is in the interior contributing to lengthen its useful life, this kind can be classified as an endophyte. These aspects make yeast a potential microbial agent able to control postharvest diseases.

In the interaction, the yeast with the phytopathogens is possible to determine different kinds of interactions such as nutrients and space competition, mycoparasitism, secretion of antibiotics, lytic enzymes, and other antifungal compounds. The importance of any one mode of action can vary between biocontrol systems (pathogen, yeast, and host).

Among all the yeasts' modes of action identified, the competition for nutrients and space is considered the most common because yeasts have the ability to grow and survive faster in the environment (host) than pathogens; thus, the bio-controlling activity is associated with an increase in the concentration of the antagonist and a decrease in the concentration of the pathogen [20]. In other cases, yeasts have the ability to adhere to fungal hyphae by restricting pathogen proliferation [21, 22], which is called parasitism and, in some cases, occurs with the production of lytic enzymes, which help bring about degradation of the cell wall of the pathogen. Other yeasts produce antibiotic compounds, case in which the control mechanism is associated with the production of secondary metabolites that inhibit the growth of pathogens [23, 24].

When studying the problem focusing on the host, it has been established that plants have the capacity to defend themselves against pathogen attacks by triggering their defense system, which can be activated by some yeasts; as a result, it is another way of action in which the yeast helps indirectly to reduce the growth and development of the pathogen.

Punja and Utkhede [25] have stated that this process can take place through the production of elicitors (signal compounds) or because of tissue colonization reducing the development of the pathogen. They have pointed out what has been reported by some researchers that the internal colonization of the tissues without causing apparent damage to the cells—characteristic associated with the endophytic yeasts—triggers the defense system of the host.

The entomologists define biocontrol like "the control of the organism by other organism," but when we talk about control of plant's diseases by yeast, the definition of biocontrol is wider because the plant's diseases are a process that involves three elements: pathogen, host, and micro environment. Then, studying the use of yeast in this context implies studying the host to and how this can change by the yeast action.

Therefore, in the case of studies on endophyte yeasts, it is necessary to characterize the host surface and its inner for establishing if it is modified and if so, define the relationship between the changes and the benefits. In regard to the production of elicitors, as a mode of action in biocontrol, this can be understood like a process in which the yeast helps the plant to activate its defense system against the attack of pathogens, however, the association between the induction of the defense system and the endophyte yeasts is not fully understood.

These aspects should also be taken into account when evaluating situations in which an endophyte yeast colonizes its host, generating in this one a different benefit from biological control. In approaching the problem from this perspective, it is possible to obtain additional information from this mutualistic relationship, which allows proposing solutions to practical problems associated with the postharvest period.

Recent investigations on the yeast *Candida guilliermondii* isolated from a heterograft tomato crop (HGTC) in Sogamoso (Boyacá, Colombia) have shown that it is able to colonize its host without generating damage to the cell walls; on the contrary, it delays loss of water; in addition, its effectiveness in biological control against *Rhizopus stolonifer* was determined [24]. These results, together with the definition of endophyte, allow us to classify this yeast as an endophyte yeast of interest in agro-industry, due to the possibility of using it in a promising way to prolong the useful life of its host.

Indeed, this endophytic yeast contributes to lengthening the useful life of its host and also can be used as an antagonist offers the possibility of using it to study this mutualistic relationship and obtain information that allows solving problems associated with the postharvest period, such as fruit quality, storage, and phytopathogen biocontrol.

However, the following questions arise: is it possible that as a result of the endophytic yeast-host interaction, changes will occur in the host? What can these changes be? Are there new modes of action of these yeasts in activating the plant defense system?

Searching for answers to these questions is possible to implement alternative methodologies that allow researchers to assess the dynamics of yeast colonization, identifying and quantifying changes in the host, and to propose another mode of action of the endophytic yeast.

4. Evaluating the action of an endophyte yeast on its host

Traditionally, to check the efficiency of a microbial antagonist and/or to evaluate a colonization process, the researchers quantify the number of microorganisms present in a plant in terms of colony forming units (CFUs). To get such measurements, it is necessary to dilute the sample, take an aliquot of it and, finally, transfer it to an appropriate medium that allows the microorganisms to grow in visible colonies [26–28].

Other investigations have proposed the direct observation of endophytic yeasts inside the plant tissues using microscopy techniques. For instance, Isaeva et al. [29] studied the distribution and species diversity of yeast in the storage tissues of fruits, seeds, and roots and found that the yeast cells were most often located in the intercellular space or in cells with intact membranes. These results suggest that internal storage tissues of plants are usually habitats of yeast and can be used as a model for studies of coevolving plant-microbe associations.

Nassar et al. [10] used light and transmission electron microscopy to observe maize root inoculated with *W. saturnus* and stained with 0.1% toluidine blue. The images show the distribution of yeast cells within the root cortex, intercellular spaces, and xylem vessels.

On the other hand, it is possible to characterize, with a vertical resolution of 10^{-9} m, the topography of fruits and vegetables from the observation of tissue samples using the atomic force microscope (AFM) [30, 31]. This methodology has also been used to evaluate the formation of antimicrobial films [32]. For their part, Isaacson et al. [33] evaluated the biomechanical properties as well as the resistance to microbial infections of tomato fruit cuticles. Because of its resolution, this microscope can be used to visualize the cell surface topography and to determine cell wall nanomechanical properties of yeast mutants [34].

In addition, evaluating the interaction of endophytic yeasts with their hosts—and taking into consideration the definition of endophyte—implicates characterizing both the surface and the interior of the host, yet it is necessary to use different methodologies from the traditional ones. From this perspective, MRI offers a non-destructive and non-invasive technique that can be used to obtain two-dimensional images of fruits and/or vegetables from which it is possible to evaluate *in vivo* changes inside, changes that take place as a result of own metabolic processes during the development and/or maturation, or associated with modifications by external agents [35–38].

According to the preceding points, the use of microscopy and MRI makes it possible to characterize qualitatively and quantitatively the host changes by the endophytic yeast action, providing information that can contribute to understanding this mutualistic relationship and to think about other conceivable action modes. Below are some of the results found when using

http://dx.doi.org/10.5772/intechopen.70326

these methodologies; for its implementation, and according to Koch's postulates, the tomato fruit was used as a host, and it was inoculated with the endophytic yeast *C. guilliermondii*.

4.1. Formation of endophytic yeast biofilms

Atomic force microscopy (AFM) enables researchers to study at a nanometric scale the distribution of endophytic yeast on the host surface as well as the topographic changes in it. Although plant tissue samples are commonly used for the implementation of this methodology, surface modifications are not only brought about by the external agent action (endophyte) but also come from the different tissues that make up the host's interior. Because of that, whole tomato fruits were used to evaluate the topography and to analyze before and after inoculation by being sprinkled with yeast *C. guilliernondii*. This methodology allows the researchers to study *in vivo* the time-related evolution of the colonization process evaluating images—taken both in contact mode and in intermittent contact mode—of the host surface.

The 3D images of the uninoculated whole fruit (zero time) surface, taken in contact mode, show that its topography is not homogeneous since it has ridges and valleys whose average value is 700 nm from the center line. It is also possible to observe bright areas associated with the epicuticular waxes, as shown in **Figure 1a**. From these images, it was determined that the average surface roughness was 240 nm.

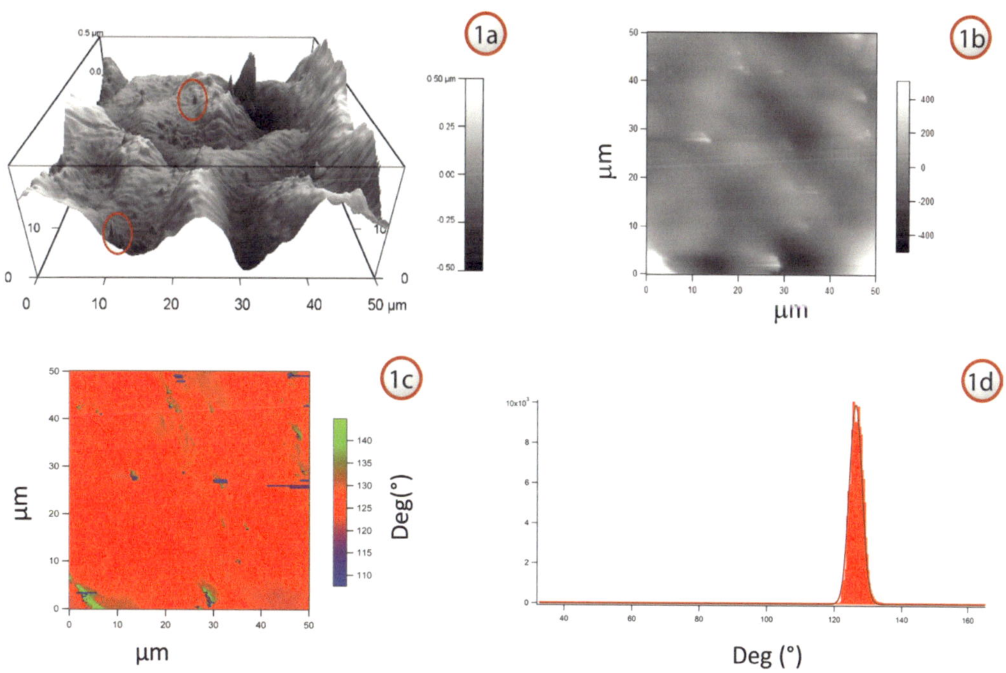

Figure 1. Images of the host surface (uninoculated fruit) obtained by AFM. (1a) 3D image taken in contact mode; the epicuticular waxes are shown in red, the vertical scale corresponds to ±0.5 μm. (1b) 2D image of the surface taken in tapping mode. (1c) Phase map. (1d) In the histogram phase for the surface of the host, there is only one phase whose value is between 120 and 135°.

The topographic characterization of the host obtained from the images taken in contact mode plus the images of the surface taken in tapping mode or intermittent contact (measuring the phase difference between the signal received when the microscope tip does not interact with the sample and the one received when the tip interacts with the sample—tap), allow to obtain information about changes in the local properties of the surface.

Figure 1b shows the two-dimensional image of the surface of the uninoculated fruit taken in tapping mode; **Figure 1c**, its corresponding map, and **Figure 1d**, its histogram phase. The results indicate that the surface has only one phase corresponding to host surface.

From the topographic images obtained 5 hours after inoculating the fruit with the yeast, it is determined that on the surface some areas associated with yeast clusters randomly appears, whose average height to the midline is 1600 nm (see **Figure 2a**). In the images of the host surface taken in tapping mode, areas of similar characteristics are observed, both in the 2D image and in the phase map (see areas surrounded by circles in **Figure 2c**).

It should be noted that the value of the phase for the yeast clusters is between 80 and 90°, a result that differs from host surface before inoculation. Additionally, the histogram phase reveals two different phases on the surface: one associated with the yeast and another associated with the surface of the fruit.

Finally, 72 hours after inoculation, the surface of the host does not present clusters as the ones described above; on the contrary, less roughness is seen, suggesting that the yeast has been colonizing and homogenizing the surface of the host (see **Figure 3a**). When calculating the roughness parameter, it is found that it has decreased to a value of 120 nm.

Concerning the map and the histogram phase, only one phase appears again, but now the value of this parameter is between 80 and 95°, for the same as the yeast clusters. This indicates that the endophyte adhered to its host formed a biofilm.

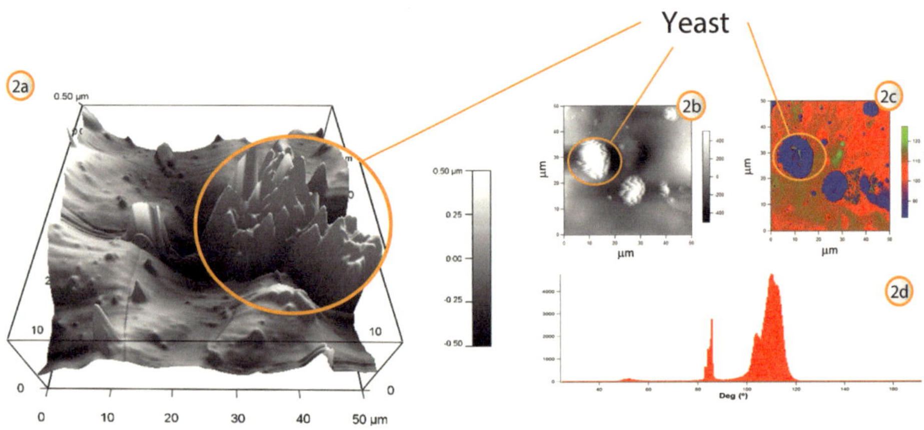

Figure 2. Host surface images taken 5 hours after inoculation with *C. guilliermondii* endophytic yeast. (2a) 3D Image taken in contact mode, the vertical scale corresponds to ±0.5 µm. (2b) 2D Image of the surface taken in tapping mode. (2c) Phase map. (2d) In the histogram phase, the two peaks confirm that the surface of the host has two phases.

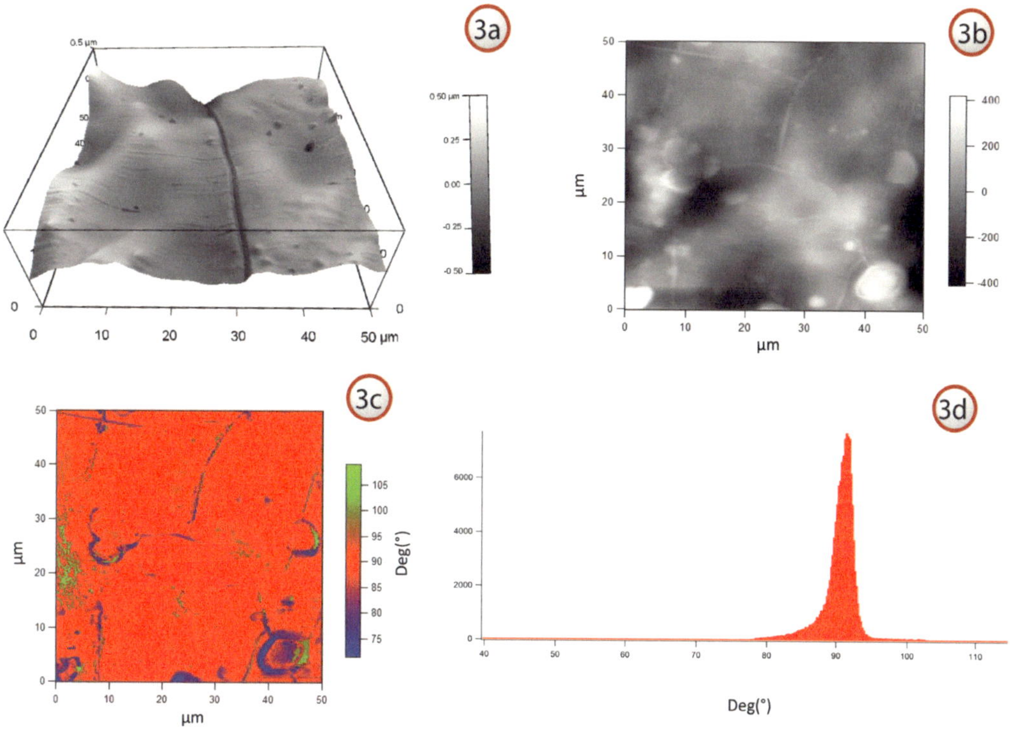

Figure 3. Host surface images obtained 72 hours after inoculation. (3a) 3D Image taken in contact mode, the vertical scale corresponds to ±0.5 μm. (3b) 2D Image of the surface taken in tapping mode. (3c) Phase map. (3d) In the histogram phase, only one phase associated with the yeast is detected.

The assessment of the host's topography allows asserting that the endophytic yeast modifies its host, reducing its surface roughness, which implies a lower adhesion of phytopathogens. In relation to the images captured in tapping mode, the results are visible how the endophytic yeast adheres to its host forming a biofilm that contributes to water retention inside the host.

4.2. Dynamics of colonization within the host

As stated by the Petrini's definition [6] "An endophyte colonizes and can live inside the living tissues of its host without causing damage," the evaluation of optical microscopy images of transverse sections of the host inoculated with the yeast enables researchers to establish if a yeast effectively is included in this definition.

In addition, this methodology allows assessing the colonization dynamics with the purpose to determine the pathways of the yeast and its average speed of migration into the host's, as well as to identify possible damage in the plant tissue and/or modifications in its structures by the endophytic action. Following the methodology proposed by Infante, Marquinez, and Moreno [39], cross-sectional images of the host can be obtained for each time after inoculation, in which the plant tissue and the yeast are simultaneously visualized, making it possible to determine the aforesaid parameters.

Figure 4 shows transverse cuts of the fruit rind inoculated with the yeast at different times after inoculation. In the control samples, the presence of endophytes is not observed. In contrast, in the inoculated samples, an increase in the number of yeasts found on the surface of the host is observed over time: in the epidermis, yeasts are observed 8 hours after the inoculation, and in the parenchyma, after 22 hours.

The images display the absence of lesions in the tissue both in the outer cuticular layer and in the cells of the epidermis and parenchyma. In relation to the yeast's pathway into the host, it is possible to establish that this endophyte, after entering, moves along the cuticular layer and then travels via apoplast, in a linear order, occupying the intercellular spaces of both the epidermis and the parenchyma as well (see **Figure 4d**).

The presence of yeast inside the host 72 hours after inoculation proves that it provides the yeast with nutrients and adequate conditions to survive, which confirms the notion of a mutualistic relationship between the endophytic yeast and the plant.

Additionally, changes by the action of the endophyte yeast in the host structures were evaluated. The results reveal an average decrease of 3 μm in the thickness of the outer cuticular layer of the bark of the tomato fruit inoculated in comparison with that of the control fruits. The outcomes are shown in **Figure 5**. The decrease in the cuticular layer thickness implies an upsurge in density, which favors the retention of water inside the fruit.

Simultaneous observation of inoculated tissues and endophyte yeasts looks into a new approach to assessing this mutual symbiosis identifying the benefits for the symbionts involved, taking into account the structural changes in the host as well as the yeasts paths and distribution patterns.

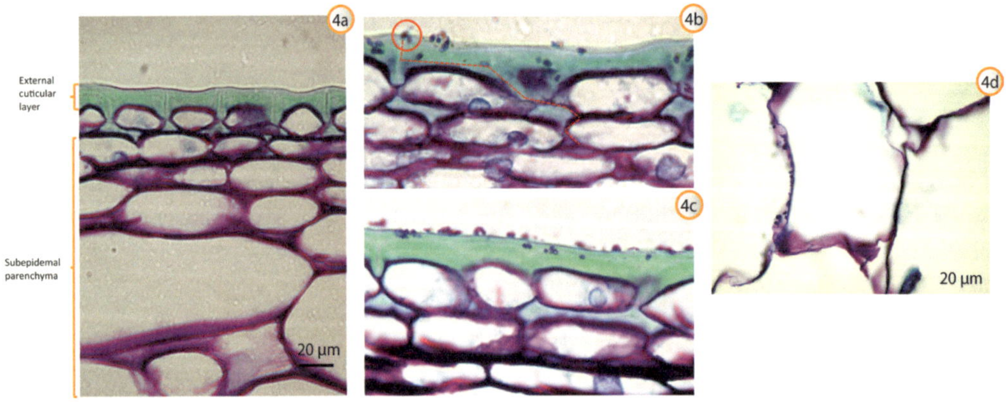

Figure 4. Cross-sectional images of tomato fruit stained with Toluidine blue, different times postinoculation. (4a) Control sample; (4b) 22 hours; (4c) 48 hours; (4d) 48 hours.

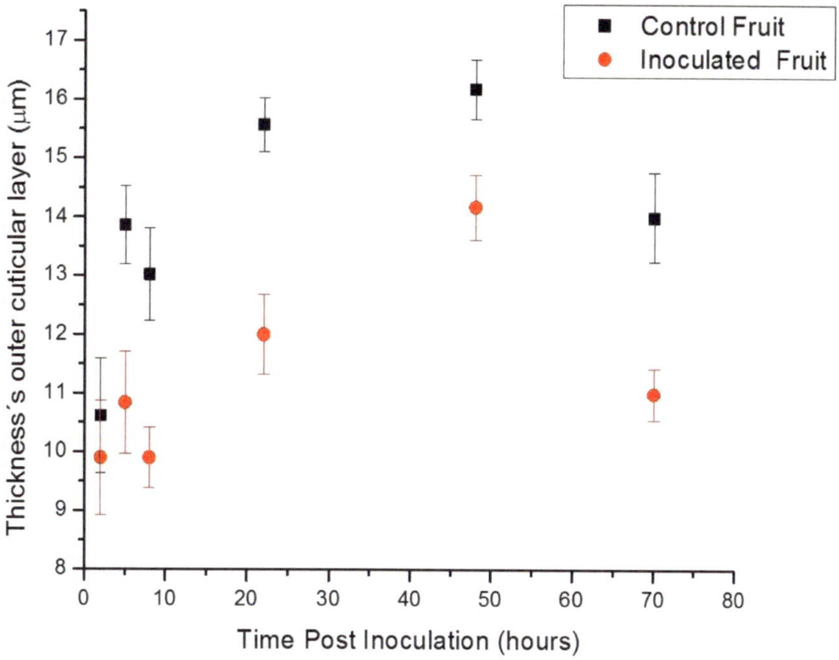

Figure 5. It measured the thickness of the outer cuticular layer of the tomato fruit rind to different times postinoculation. The differences in thickness between the control fruits and the inoculated ones are statistically significant.

4.3. How an endophytic yeast modifies the interior of its host

The results reported in relation to the yeast *C. guilliermondii* have shown that it adheres to the host forming a biofilm and colonizes its interior without causing damage to the cell walls. Instead, it contributes to decreasing both the phytopathogens attacks and the water loss. As this is an endophyte yeast, it is interesting to identify changes in the internal structures of the host and its relation to the benefits that it receives, with the intention of deepening the knowledge of this symbiosis.

To study these alterations, it is advisable to use magnetic resonance imaging (MRI)—a non-invasive technique—which enables investigators to see changes *in vivo* inside the host triggered by the endophyte's action, as in the case of the modifications that happened in the tomato fruit inoculated with *C. guilliermondii*. On the minus side, MRI does not permit researchers to observe simultaneously the host and the yeast—unlike the techniques of MRI microscopy—since in this case, the scale resolution is the tenth of a millimeter.

The main advantage of this technique is the possibility to obtain images weighted by different parameters—relaxation times (T2), proton density, and diffusion, among others—which correspond to the characteristics of the evaluated system. With the aim to see the temporal evolution of the host, images of tomato fruits inoculated by sprinkling with the yeast *C. guilliermondii* were taken at different times after inoculation.

Changes in the dimensions of the host were evaluated. The results obtained indicate that the most affected fruit region by the yeast is the pericarp; also, the diameter of the inoculated fruits decreases more slowly; however, the pericarp thickness diminishes more in comparison with the control fruits (**Figure 6**). This suggests that there are structural changes by the action of the endophyte in this region of the fruit, which can contribute to water retention and, as a consequence, delay the loss of turgor. This is the reason why the decrease of its size is slower compared with the control fruits. Nevertheless, it is necessary to evaluate parameters such as relaxation time (T2) and mobility to confirm these assertions.

With the propose of establishing the biochemical changes within the host, T2-weighted images were taken; the results indicate differences in the values associated with this parameter for the different regions of the fruit (see **Figure 7**). It was also found that T2 decreases in both control and inoculated fruits, signifying molecular variations associated with postharvest processes. However, this decrease occurs in the inoculated fruits more rapidly, which evidences lessening of mobility due to molecular modifications inside the fruit.

Finally, the diffusion-weighted images allow establishing changes in the mobility of molecules, which is a fundamental aspect in this case because the yeast helps to retain water inside the host. From the obtained images, the apparent diffusion coefficient (ADC) was calculated. It is lower in the pericarp region of the inoculated fruits than for the control ones, which indicates that the host is modified by the action of the endophyte, reducing the movement of the water molecules inside. This result, combined with that reported for the T2 parameter, allows to state that in the fruits inoculated with the yeast the water molecules present in the pericarp region are surrounded by different molecules that limit their mobility.

Evaluating the images obtained by MRI, it is possible to sustain that the endophytic yeast modifies the interior of the host; in the case of the inoculated tomato fruit, a decrease in its thickness was observed for the pericarp region in comparison with the control fruits, fact that

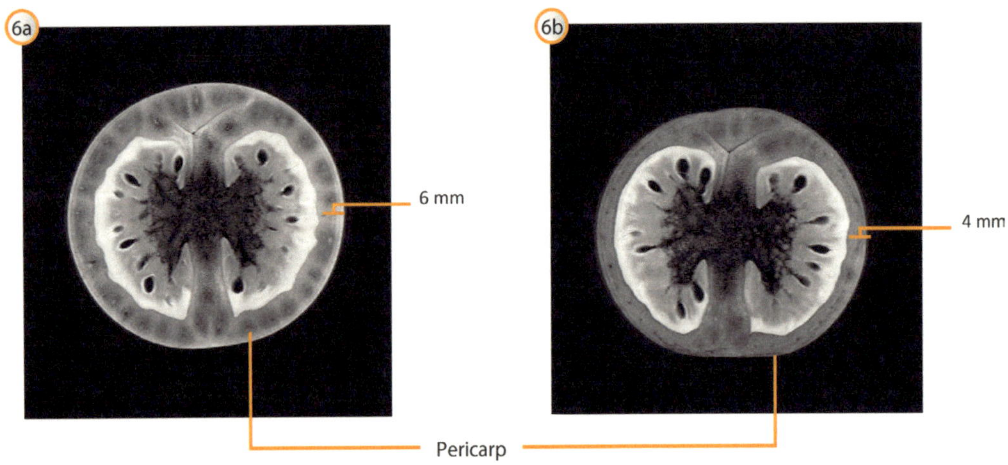

Figure 6. High-resolution images of a cross-section of the inoculated tomato fruit. (6a) Zero time. (6b) 14 days postinoculation.

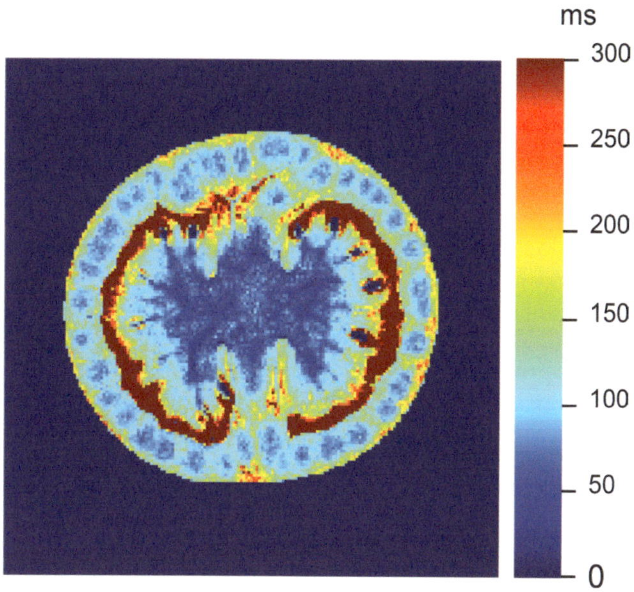

Figure 7. T2 map in a cross-section of the fruit. High values of T2 (more than 250 ms) specify zones with water molecules that can move easily; on the contrary, low values (70 ms) indicate the presence of different molecules.

correlates with biochemical changes that help to reduce the mobility of the molecules in this region. These aspects together favor the retention of water inside the host contributing to maintaining the quality of the fruit.

5. Another approach on endophytic yeasts' action

Reported research has shown that endophytic yeasts can be used in different agro-industrial applications contributing to host and/or pathogen control improvements, however, some aspects remain unclear. For instance, the way the yeast triggers the defense system in the host, where the relationship between the elicitors and the antagonist provides a field to be explored. Another aspect that has drawn attention is the formation of biofilms and how these can be used to improve biological control [16]; additionally, it is necessary to evaluate the changes produced in the host by the yeast's action and its incidence. All of them are topics that to date have been little explored.

The relationships established between yeast, pathogen, host, and metabolic changes that occur in the host during the postharvest period allow to understand the plant-endophyte mutualistic association and define other modes of action.

Evaluating these relationships focusing on the host, it was found that the metabolic processes associated with the postharvest period—such as starch degradation, water loss, and disassembly of cell walls—lead to changes that affect the quality of the product. Concerning its interactions with pathogens, these colonize the host generating various diseases, to which the host can respond by activating its defense system and producing antifungal compounds. On the contrary, their relation with the endophytic yeasts is of mutualistic character, since these generate a benefit for the host while it offers to them optimal conditions for their survival.

The relationships described above are shown in **Figure 8**; the arrows indicate direct interactions; however, when it comes to endophytic yeasts, it is necessary to consider indirect relationships, in which the yeast can modify its host generating benefits in it, helping solve some of the postharvest period problems.

In the previous section, the results obtained when evaluating changes in the host (tomato fruit) by the action of the endophyte yeast (*C. guilliermondii*) were presented. Using atomic force microscopy (AFM), it could be established that the surface roughness of the inoculated host diminishes when a yeast biofilm is formed, besides it contributes to retaining the water inside the host prolonging its useful life.

On the other hand, when the samples inoculated with the yeast were evaluated by optical microscopy (OM), it was determined that the thickness of the outer cuticle layer showed an average decrease of 3 μm in comparison with the control samples, suggesting an increase in the density of the same and, therefore, changes in its permeability.

It should be noted that in relation to cuticle evaluation and its function in resistance to phytopathogens, Curvers et al. [40] studied a mutant of tomato (*Solanum lycopersicum*) with reduced abscisic acid (ABA) production, and established that it presents increased resistance to the necrotrophic fungus *Botrytis cinerea*. They further compared the thickness of the cuticle of the mutant with that of other evaluated tomato fruits, identifying that the cuticular layer of the first one presents a decrease in the thickness, which favors the signaling processes.

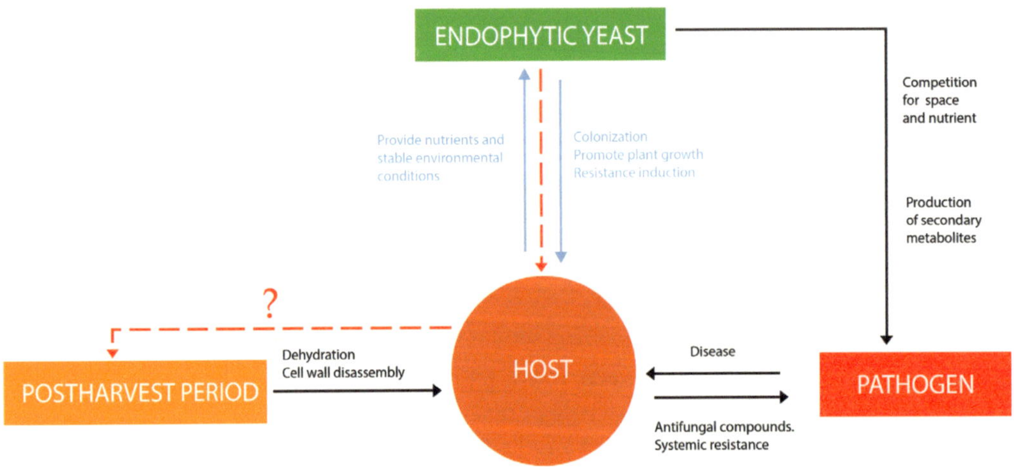

Figure 8. Interactions between host, yeast, pathogen, and postharvest processes. The blue arrow indicates a yeast–host mutualistic relationship; the question mark points to a possible indirect relationship between the yeast and postharvest period.

Previous studies about yeast *C. guilliermondii* determined its effectiveness in the control of *Rhizopus stolonifer* and its production of secondary metabolites; this outcome together with the results found by means of microscopy techniques allow to affirm that the effectiveness of an endophyte yeast in the biological control could be associated with more than one mode of action, one of which may be related to structural changes in the host by action of the endophyte.

Lastly, from MRI, it was determined that with respect to the control fruits, in the fruits inoculated with the endophytic yeast appear a decrease both in the thickness of the pericarp and in the mobility of the molecules present in this region of the fruit; changes that favor the retention of water inside. **Figure 9** shows the modifications generated in the different structures of the host by the action of yeast and its relation to the observed benefits.

According to the abovementioned determination, it is possible to highlight several aspects that contribute to deepening the knowledge of endophyte yeasts and their use in the search for solutions to problems typical of the postharvest period.

The first one refers to the fact that the endophyte yeast colonizes not only the surface of the host but also enters into it and remains inside it without causing damage: evidence of the mutualistic relationship between the symbionts.

In addition, from the results found, it is possible to propose another mode of action of the endophytic yeasts: they generate propitious structural changes in the surface and the interior of the host, which reduce phytopathogen attacks and loss of water. Therefore, it can be said that the endophytic yeasts could be used to help solve some of the problems relevant to agro-industry.

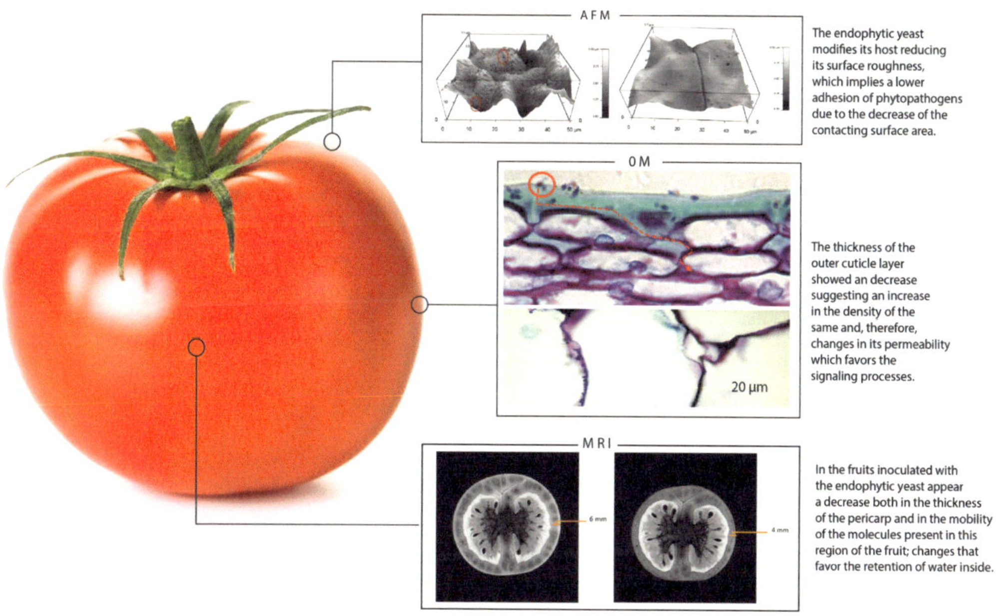

Figure 9. Physical modifications in the host (tomato fruit) by the action of the endophyte yeast (*C. guilliermondii*).

It is noteworthy that this mutualistic coexistence of plant-endophytic yeast can be applied to develop healthy and friendly alternatives that are advantageous to the environment, offering organic food to the consumers and avoiding the use of agrochemicals and genetic engineering intended to enhance the quality of fruits and vegetables.

6. Conclusion

This chapter shows a new way to understand the endophytic yeasts, analyzing variations in their host looked through microscopy and the magnetic resonance imaging. The results confirmed the Petrini's definition: "An endophyte colonizes and can live inside the living tissues of its host without causing damage" additionally —observing the inoculated host— it is thinkable to propose a new mode of yeast action in which the physical characteristics of the surface and the inside of the host change by the action of the yeast, contributing to improve their quality during the postharvest period, without causing health problems to the humans beings, because by this way the use of chemicals to control phytopathogens is avoided.

The new information about endophytic yeast opens the possibility to new researches: how the host "understand" that this microorganism is good for it?; how is the process in the host that allows the entry of the endophytic yeast?; how can this kind of yeast be used to obtain organic products in order to improve the health?; how does the biochemical environment of the host changes by the yeast?

I hope that these new methodologies and information about the endophytic yeast contribute to solve these questions.

Author details

Esperanza del Pilar Infante Luna

Address all correspondence to: epinfantel@udistrital.edu.co

Universidad Distrital Francisco José de Caldas, Bogotá, Colombia

References

[1] Sharma RR, Singh D, Singh R. Biological control of postharvest diseases of fruits and vegetables by microbial antagonists: A review. Biological Control. September 2009;**50**(3):205-221. [Internet]. Available from: http://linkinghub.elsevier.com/retrieve/pii/S1049964409001236 [Accessed: November 6, 2012]

[2] Isaeva OV, Glushakova AM, Garbuz SA, Kachalkin AV, Chernov IY. Endophytic yeast fungi in plant storage tissues. Biology Bulletin. February 9, 2010;**37**(1):26-34. [Internet]

Available from: http://www.springerlink.com/index/10.1134/S1062359010010048 [Accessed: January 20, 2013]

[3] Schulz B, Boyle C, Sieber TN, editors. Microbial Root Endophytes. Berlin: Springer-Verlag; 2006. p. 367

[4] Hirsch G, Braun U 1992. Communities of parasitic microfungi. In: Winterhoff W, ed. Handbook of vegetative science, Volume 19. Fungi in vegetation science. Dordrecht, the Netherlands: Kluwer Academic Publishers, 225-250

[5] Stone J, Bacon C, White J. An overview of endophytic microbes: Endophytism defined. In: Bacon C, White J, editors. Microbial Endophytes. New York, USA: Marcel Der; 2000

[6] Petrini O. Fungal endophytes of tree leaves. In: Andrews JH, Hirane SS, editors. Micorbial Ecology of Leaves. New York, USA: Springer; 1991. pp. 179-197

[7] Xin G, Glawe D, Doty S. Characterization of three endophytic, indole-3-acetic acid producing yeast ocurring in populus trees. Mycological Research. 2009;**113**:973-980

[8] Dal Bello G, Mónaco C, Rollan MC, Lampugnani G, Arteta N, Abramoff C, et al. Biocontrol of postharvest grey mould on tomato by yeasts. Journal of Phytopathology. May 22, 2008;**156**(5):257-263. [Internet] Available from: http://doi.wiley.com/10.1111/j.1439-0434.2007.01351.x [Accessed: March 1, 2014]

[9] Bautista-Rosales PU, Calderon-Santoyo M, Servín-Villegas R, Ochoa-Álvarez NA, Ragazzo-Sánchez JA. Action mechanisms of the yeast *Meyerozyma caribbica* for the control of the phytopathogen *Colletotrichum gloeosporioides* in mangoes. Biological Control. Jun 2013;**65**(3):293-301. [Internet] Available from: http://linkinghub.elsevier.com/retrieve/pii/S1049964413000595 [Accessed: September 14, 2014]

[10] Nassar AH, El-Tarabily KA, Sivasithamparam K. Promotion of plant growth by an auxin-producing isolate of the yeast *Williopsis saturnus* endophytic in maize (Zea mays L.) roots. Biology and Fertility of Soils. July 1, 2005;**42**(2):97-108. [Internet] Available from: http://www.springerlink.com/index/10.1007/s00374-005-0008-y [Accessed: November 30, 2012]

[11] Zhao Y, Wang R, Tu K, Liu K. Efficacy of preharvest spraying with *Pichia guilliermondii* on postharvest decay and quality of cherry tomato fruit during storage. African Journal of Biotechnology. 2011;**10**(47):9613-9622

[12] De Bary A. 1879. Die Erscheinung der Symbiose. In: Trubner KJ, ed. Vortrag auf der Versammlung der Naturforscher und Ärtze zu Cassel. Strassburg, Germany: Verlag, 1-30

[13] Singh LP, Gill SS, Tuteja N. Unraveling the role of fungal symbionts in plant abiotic stress tolerance. Plant Signaling & Behavior. February 2011;**6**(2):175-191

[14] Sieber T. Endophytic fungi in forest trees: Are they mutualists? Fungal Biology Reviews. 2007;**21**:75-89

[15] Nunes CA. Biological control of postharvest diseases of fruit. European Journal of Plant Pathology. December 21, 2012;**133**(1):181-196. [Internet] Available from: http://link.springer.com/10.1007/s10658-011-9919-7 [Accessed: October 14, 2014]

[16] Liu J, Sui Y, Wisniewski M, Droby S, Liu Y. Review: Utilization of antagonistic yeasts to manage postharvest fungal diseases of fruit. International Journal of Food Microbiology. October 15, 2013;**167**(2):153-160. [Internet] Available from: http://www.ncbi.nlm.nih.gov/pubmed/24135671 [Accessed: October 14, 2014]

[17] Robiglio A, Sosa MC, Lutz MC, Lopes CA, Sangorrín MP. Yeast biocontrol of fungal spoilage of pears stored at low temperature. International Journal of Food Microbiology. June 30, 2011;**147**(3):211-216. [Internet] Available from: http://www.ncbi.nlm.nih.gov/pubmed/21546110 [Accessed: September 23, 2014]

[18] Lai T, Wang Y, Li B, Qin G, Tian S. Defense responses of tomato fruit to exogenous nitric oxide during postharvest storage. Postharvest Biology and Technology. November 2011;**62**(2):127-132. [Internet] Available from: http://linkinghub.elsevier.com/retrieve/pii/S0925521411001281 [Accessed: July 22, 2014]

[19] Jiang F, Chen J, Miao Y, Krupinska K, Zheng X. Identification of differentially expressed genes from cherry tomato fruit (*Lycopersicon esculentum*) after application of the biological control yeast *Cryptococcus laurentii*. Postharvest Biology and Technology. September 2009;**53**(3):131-137. [Internet] Available from: http://linkinghub.elsevier.com/retrieve/pii/S0925521409000660 [Accessed: January 20, 2013]

[20] Wang Y, Yu T, Xia J, Yu D, Wang J, Zheng X. Biocontrol of postharvest gray mold of cherry tomatoes with the marine yeast *Rhodosporidium paludigenum*. Biological Control. May 2010;**53**(2):178-182. [Internet] Available from: http://linkinghub.elsevier.com/retrieve/pii/S1049964410000101 [Accessed: January 20, 2013]

[21] Chan Z, Tian S. Interaction of antagonistic yeasts against postharvest pathogens of apple fruit and possible mode of action. Postharvest Biology and Technology. May 2005;**36**(2):215-223. [Internet] Available from: http://linkinghub.elsevier.com/retrieve/pii/S0925521405000128 [Accessed: January 20, 2013]

[22] Lima JR, Gondim DMF, Oliveira JTA., Oliveira FSA, Gonçalves LRB, Viana FMP. Use of killer yeast in the management of postharvest papaya anthracnose. Postharvest Biology and Technology. September 2013;**83**:58-64. [Internet] Available from: http://linkinghub.elsevier.com/retrieve/pii/S0925521413000884 [Accessed: July 22, 2014]

[23] Mari M, Martini C, Spadoni A, Rouissi W, Bertolini P. Biocontrol of apple postharvest decay by *Aureobasidium pullulans*. Postharvest Biology and Technology. November 2012;**73**:56-62. [Internet] Available from: http://linkinghub.elsevier.com/retrieve/pii/S0925521412001287 [Accessed: October 14, 2014]

[24] Celis ZC, Moreno DG, Sequeda-castañeda LG, García A, Albarracín DM, Charry B, Claudia L. Determining the effectiveness of *Candida guilliermondii* in the biological control of *Rhizopus stolonifer* in postharvest tomatoes. Universitas Scientiarum. 2014;**19**(1): 51-62

[25] Punja ZK, Utkhede RS. Using fungi and yeasts to manage vegetable crop diseases. Trends in Biotechnology. September 2003;**21**(9):400-407. [Internet] Available from: http://www.ncbi.nlm.nih.gov/pubmed/12948673 [Accessed: October 4, 2014]

[26] Ahmed L, Martin-Diana AB, Rico D, Barry-Ryan C. Effect of delactosed whey permeate treatment on physico-chemical, sensorial, nutritional and microbial properties of whole tomatoes during postharvest storage. LWT – Food Science and Technology. April 2013;**51**(1):367-374. [Internet] Available from: http://linkinghub.elsevier.com/retrieve/pii/S002364381200429X [Accessed: March 1, 2014]

[27] Mahmoud BSM. The effects of X-ray radiation on Escherichia coli O157:H7, *Listeria monocytogenes*, *Salmonella enterica* and *Shigella flexneri* inoculated on whole Roma tomatoes. Food Microbiology. December 2010;**27**(8):1057-1063. [Internet] Available from: http://www.ncbi.nlm.nih.gov/pubmed/20832685 [Accessed: March 1, 2014]

[28] Adekunte A, Tiwari B, Cullen P, Scannell AG, O'Donnell C. Effect of sonication on colour, ascorbic acid and yeast inactivation in tomato juice. Food Chemistry. October 2010;**122**(3):500-507. [Internet] Available from: http://linkinghub.elsevier.com/retrieve/pii/S0308814610001068 [Accessed: March 1, 2014]

[29] Isaeva OV, Glushakova AM, Yurkov AM, Chernov IY. The yeast *Candida railenensis* in the fruits of English oak (*Quercus robur* L.). Microbiology. June 7, 2009;**78**(3):355-359. [Internet] Available from: http://www.springerlink.com/index/10.1134/S002626170903014X [Accessed: January 20, 2013]

[30] Wiśniewska SK, Nalaskowski J, Witka-Jeżewska E, Hupka J, Miller JD. Surface properties of barley straw. Colloids and Surfaces B: Biointerfaces. June 2003;**29**(2-3):131-142. [Internet] Available from: http://linkinghub.elsevier.com/retrieve/pii/S0927776502001789 [Accessed: May 9, 2013]

[31] Xin Y, Chen F, Yang H, Zhang P, Deng Y, Yang B. Morphology, profile and role of chelate-soluble pectin on tomato properties during ripening. Food Chemistry. July 2010;**121**(2):372-380. [Internet] Available from: http://linkinghub.elsevier.com/retrieve/pii/S0308814609014587 [Accessed: December 6, 2012]

[32] La Storia A, Ercolini D, Marinello F, Mauriello G. Characterization of bacteriocin-coated antimicrobial polyethylene films by atomic force microscopy. Journal of Food Science. May 2008;**73**(4):T48-T54. [Internet] Available from: http://www.ncbi.nlm.nih.gov/pubmed/18460145 [Accessed: March 1, 2014]

[33] Isaacson T, Kosma DK, Matas AJ, Buda GJ, He Y, Yu B, et al. Cutin deficiency in the tomato fruit cuticle consistently affects resistance to microbial infection and biomechanical properties, but not transpirational water loss. The Plant Journal. October 2009;**60**(2):363-377. [Internet] Available from: http://www.ncbi.nlm.nih.gov/pubmed/19594708 [Accessed: January 7, 2013]

[34] Dague E, Bitar R, Ranchon H, Durand F, Fran JM. An atomic force microscopy analysis of yeast mutants defective in cell wall architecture. Yeast. 2010;**27**:673-684

[35] Galed G, Fernández-Valle ME, Martínez A, Heras A. Application of MRI to monitor the process of ripening and decay in citrus treated with chitosan solutions. Magnetic Resonance Imaging. January 2004;**22**(1):127-137. [Internet] Available from: http://www.ncbi.nlm.nih.gov/pubmed/14972402 [Accessed: September 23, 2014]

[36] Musse M, Quellec S, Devaux M-F, Cambert M, Lahaye M, Mariette F. An investigation of the structural aspects of the tomato fruit by means of quantitative nuclear magnetic resonance imaging. Magnetic Resonance Imaging. June 2009;**27**(5):709-719. [Internet] Available from: http://www.ncbi.nlm.nih.gov/pubmed/19144488 [Accessed: January 20, 2013]

[37] Zhang L, McCarthy MJ. Measurement and evaluation of tomato maturity using magnetic resonance imaging. Postharvest Biology and Technology. May 2012;**67**:37-43. [Internet] Available from: http://linkinghub.elsevier.com/retrieve/pii/S0925521411002912 [Accessed: September 5, 2014]

[38] Defraeye T, Lehmann V, Gross D, Holat C, Herremans E, Verboven P, et al. Application of MRI for tissue characterisation of "Braeburn" apple. Postharvest Biology and Technology. January 2013;**75**:96-105. [Internet] Available from: http://linkinghub.elsevier.com/retrieve/pii/S0925521412001913 [Accessed: September 29, 2014]

[39] Infante E, Marquinez X, Moreno G. Tomato peel (*Solanum lycopersicum* L.) colonization by the endophyte yeast *Candida guilliermondii* (Castellani) Langeron et Guerra. Agronomía Colombiana 2012;**30**(3):388-394

[40] Curvers K, Seifi H, Mouille G, de Rycke R, Asselbergh B, Van Hecke A, et al. Abscisic acid deficiency causes changes in cuticle permeability and pectin composition that influence tomato resistance to *Botrytis cinerea*. Plant Physiology. October 2010;**154**(2):847-860. [Internet] Available from: http://www.pubmedcentral.nih.gov/articlerender.fcgi?artid=2949027&tool=pmcentrez&rendertype=abstract [Accessed: July 22, 2014]